PIPELINE ENGINEERING COMPUTATIONS

HANDBOOK OF PIPELINE ENGINEERING COMPUTATIONS

$P=Q(x)=(b_1t+b_2t$

b_3t^3 b_4t^4 $Q\ (-X$

$e(x)^2/_2$

$x10$ $\frac{1}{\sqrt{2\pi}}$

By ALEX MARKS
Marks Research Consultants
Calgary, Alberta Canada

PPC BOOKS
Petroleum Publishing Company
Box 1260
Tulsa, Oklahoma 74101

To Susan and David

PPC Books Division, The Petroleum Publishing Company
1421 South Sheridan Road
P. O. Box 1260
Tulsa, Oklahoma 74101

International Standard Book Number: 0-87814-094-8

Printed in the U.S.A.

1 2 3 4 5 -- 83 82 81 80 79

Preface

It was the author's purpose to write a handbook for pipeline engineers that would be a reference book of practical, how-to-do-it information.

A pipeline engineer with a problem will be able to turn to this handbook and find the help he needs to solve that problem. But in addition to this, programmable calculator software is added to speed the computations and provide portability at the same time.

The field of pipeline engineering has been growing so rapidly that all pipeline engineers are faced with the problem of keeping up to date. In order to grow with their profession, they will need to extend their interests and skills and even a casual reading of the material included in this handbook will help them to move ahead in the right direction.

This handbook is very practical and each section is designed to facilitate the understanding of the material. Each section also enables the reader to acquire confidence in his mastery and use of the calculator and its software.

Finally, I wish to acknowledge my wife Susan Marks, who as a capable mathematician, painstakingly checked and verified the software in this handbook.

A. Marks
Calgary, Alberta
August 1978

INTRODUCTION

It has long been obvious that a handbook of pipeline engineering computations has been needed in the pipe-lining industry.

The author has attempted to present a pragmatic approach to these computations by utilizing programmable calculator programs. The handbook is divided into three principal parts.

Part One presents material that is to be of use to the pipeline engineer engaged in feasibility studies.

Part Two depicts material that is to be of use to the pipeline engineer engaged in detailed design.

Part Three directs itself to material of use to the pipe-line construction engineer.

The three principal parts are programmed for a turnkey approach to pipeline engineering problems with the use of a Hewlett-Packard HP-67 for execution on a portable basis. All programs are documented for theoretical verification.

The handbook is intended to give a review of the most important computations in pipeline engineering. This field, of course, is an extremely specialized one, and it would be unrealistic, not to say foolhardy, to attempt a complete and balanced review of the current state of knowledge.

The accumulation of data for this handbook and programs was begun while the author was project evaluation engineer for a crude gathering company in the United States, and was continued while he saw service in the Middle East, Indonesia, Venezuela and Canada.

Alex Marks
Calgary, Alberta,
Canada
August 1977

TABLE OF CONTENTS

SECTION ONE

FEASIBILITY

1.0 Diameter for Minimum Investment

1.1 Introduction

This program selects the diameter for the minimum investment utilizing the calculus principle of maxima and minima.

1.2 Theoretical Formulae

$$D = (C_2/C_1\ \eta\ (1.906 \times 10^{-5})(.0018 + .03z^{.355})(Q)^{2.65}(S)\ )^{.177}$$

WHERE:

D = I.D. - inches (minimum)
z = viscosity in c.s.
Q = flowrate - bpd
S = specific gravity
C_2 = unit cost of pumping facilities - \$/HP - does not include tankage
C_1 = unit installation costs - \$/inch I.D./mile. Includes purchase, transportation, and installation
η = pump efficiency

Other formulae available for differentiation are:

Name of Formula	Service	Formula
D'Arcy-Weisbach	Crude	$Q = 4.064 \times \frac{D^{2.5}\ P^{.5}}{S^{.5}\ f^{.5}}$ Where f is expressed by the Colebrook-White equation: $\sqrt{f} = \frac{1}{-2\ \log_{10}(e/3.7D + 2.51/N_R\sqrt{f})}$ $N_R = \frac{92.24Q}{Dv}$
Hazen-Williams	Products	$Q = \frac{0.148D^{2.63}P^{.54}c}{S^{.54}}$
Miller's	Products	$Q = \frac{4.06D^{2.5}P^{.5}}{S^{.5}}\ \ln\ (D^3SP/\mu^2 + 4.35)$
T.R. Aude	Products	$Q = \frac{20.9D^{2.66}P^{.552}K}{S^{.448}\ \mu^{.104}}$
Service Pipe Line	Crude	$Q = \frac{18P^{.572}D^{2.716}}{v^{.144}\ S^{.572}}$

Where f in the D'Arcy-Weisbach equation was said to equal $\frac{0.3305}{N_R^{.252}}$.
When this value of f was substituted into the D'Arcy-equation, the Service Pipe Line equation became the result.

WHERE:

Q = rate, in barrels of 42 U.S. gallons per day
D = inside diameter of pipe, in inches
P = pressure drop, in pounds per square inch per mile
S = specific gravity of liquid
f = D'Arcy-Weisbach friction factor
e = height of protuberances on inside pipe wall, in inches
N_R = Reynolds number
v = kinematic viscosity, in centistokes
c = Hazen-Williams friction factor
μ = absolute viscosity, in centipoises
K = T.R. Aude's friction factor, pipe efficiency

Note that this program does not take into account operating expenses.

Annual cost parameters can be estimated from the midyear - Oil & Gas Journal. Pipeline Economics Publication.

LP-F-001 - Preliminary Selection of Diameter for Minimum Investment

Input:

(1) U.S.$/BHP installed - $
(2) U.S.$/inch I.D./mi. - installed - $
(3) pumping efficiency - n
(4) specific gravity of fluid - n
(5) viscosity of fluid - centistokes
(6) flowrate of fluid - bpd

Computes:

(1) diameter for minimum investment

STEP	INSTRUCTIONS	KEYS	OUTPUT UNITS
1.	load card LP-F-001, side 1		
2.	input U.S. dollars per installed horsepower	A	$
3.	input U.S. dollars per installed pipeline - I.D. - mile	A	$
4.	input pumping efficiency	A	n
5.	input specific gravity	A	n
6.	input fluid viscosity at operating temperature	A	c.s.
7.	input flowrate	A	bpd
8.	Compute: DFMI	B	inches
9.	Optional:		
	(1) input a new cost of $/HP installed and compute DFMI - all else constant	C	inches
	(2) input a new cost of $/inch I.D./mi. installed and compute DFMI - all else constant	D	inches
	(3) input a new flowrate of bpd	E	inches
	(4) convert gpm to bpd	fe	bpd
10.	To clear and reset for new case	fa	

Example: LP-F-001 - Preliminary Selection of Diameter for Minimum Investment

Given Conditions:

$/HP installed: 350.00
$/inch I.D./mi.: 10,000.00
pumping efficiency: 0.77
specific gravity: 0.585
viscosity: 0.31 c.s.
flowrate: 52,910 bpd

Execution: Load card LP-F-001, side 1

Store parameters in this order:

KEYSTROKES	OUTPUT
350 "A"	350.000
10,000 "A"	10,000.000
.77 "A"	.770
.585 "A"	.585
.31 "A"	.310
52,910 "A"	52,910.000

To Compute:

Press "B" ____________ 6.40 inches - DFMI

New $/HP - $500/HP

500 "C" ____________ 6.82 inches - DFMI

New $/ inch I.D./mi. - $20,000

20,000 "D" ____________ 6.03 inches - DFMI

New Flowrate - 35,000 bpd

35,000 "E" ____________ 4.97 inches - DFMI

Option to Convert gpm to bpd: 700 gpm to bpd

700 "fe" ____________ 24,000.02 bpd

"fa" ____________ reset for new case

STEP	KEY ENTRY	KEY CODE
001	f LBL A	31 25 11
	STO (i)	33 24
	f GSB 1	31 22 01
	h RTN	35 22
	f LBL 1	31 25 01
	f ISZ	31 34
	R/S	84
	h ST I	35 33
	GTO 1	22 01
010	h RTN	35 22
	f LBL B	31 25 12
	RCL 0	34 00
	RCL 1	34 01
	÷	81
	RCL 2	34 02
	÷	81
	.	83
	0	00
	0	00
020	0	00
	0	00
	1	01
	9	09
	0	00
	6	06
	x	71
	RCL 3	34 03
	x	71
	STO A	33 11
030	RCL 4	34 04
	.	83
	3	03
	5	05
	5	05
	h y^x	35 63
	.	83
	0	00
	3	03
	x	71
040	.	83
	0	00
	0	00
	1	01
	8	08
	+	61
	RCL A	34 11
	x	71
	STO A	33 11
	RCL 5	34 05
050	2	02
	.	83
	6	06
	5	05
	h y^x	35 63
	RCL A	34 11
	x	71
	.	83
	1	01
	7	07
060	7	07
	h y^x	35 63
	STO A	33 11
	h RTN	35 22
	f LBL E	31 25 15
	STO 5	33 05
	GTO B	22 12
	h RTN	35 22
	f LBL C	31 25 13
	STO 0	33 00
070	GTO B	22 12
	h RTN	35 22
	f LBL D	31 25 14
	STO 1	33 01
	GTO B	22 12
	h RTN	35 22
	g LBLf a	32 25 11
	f CL REG	31 43
	h RTN	35 22
	g LBLf e	32 25 15
080	3	03
	4	04
	.	83
	2	02
	8	08
	6	06
	x	71
	STO 5	33 05
	h RTN	35 22

References

1-1 Downs, G.F., "Formulas for selecting Pipe-line Diameter for Minimum Investment". Division of Transportation, 33rd annual meeting, A.P.I., November 10, 1953.

1-2 Oil & Gas Journal, Annual Cost Parameters, August 23, 1976. Page 77 to 103. "Pipeline Economics".

2.0 Minimum Wall Thickness/Design Pressure Computation - Barlow's Formula

2.1 Introduction

This program utilizes the Barlow Formula to obtain design parameters.

2.2 Theoretical Formulae

$$P = \frac{2\ st \times F \times E \times T}{D}$$

Where:

P = design pressure, psig
s = specified minimum yield strength, PSI
D = nominal outside diameter - inches
t = nominal wall thickness - inches
F = construction type design factor
E = longitudinal joint factor
T = temperature derating factor

2.3 Construction Type Design Factor (ANSI B31.8)

Type A	F =	.72 for cross country locations etc. Includes crossings in casings
Type B	F =	.60 for fringe areas around cities and towns,etc. Includes road crossings without casings and crossings in casings if in type 2 locations.
Type C	F =	.50 for commercial and residential sections etc.
Type D	F =	.40 for areas with multi story buildings etc.

2.4 Longitudinal Joint Factor

	E
Seamless & Welded pipe	1.0
Fusion welded A134 & A139	.8
Spiral welded A211	.8
Butt Welded ASTM-A53,API 5L	.6

2.5 Temperature Derating Factor

°F	T
250 or less	1.0
300	.967
350	.933

LP-F-002 - Minimum Wall Thickness/Design Pressure Computation Barlow's Formula

Input:

(1) outside diameter of pipeline - inches
(2) specified minimum yield of pipe - PSI
(3) construction type design factor - n

Input/Compute:

(1) minimum wall thickness for given design pressure - inches
(2) design pressure for given wall thickness

STEP	INSTRUCTIONS	KEYS	OUTPUT UNITS
1.	load card LP-F-002, side 1		
2.	input outside diameter of pipeline	C	inches
3.	input specified minimum yield of pipe	D	PSI
4.	input construction type design factor	E	n
5.	input/compute minimum wall thickness for a given design pressure	A	inches
6.	input/compute design pressure for a given pressure	B	PSIG
7.	optional: clear and reset for new case	fa	

Example: LP-F-002 - Wall Thickness/Design Pressure Computation

Given Conditions:

pipeline outside diameter: 18 inches
specified minimum yield: 70,000 PSI
construction type design factor: .72
design pressure: 1440 psig

Execution: Load card LP-F-002, side 1

Store parameters in this order:

KEYSTROKES	OUTPUT
18 "C"	18.00
70,000 "D"	70,000.00
.72 "E"	.72
1440 "B"	1,440.00

To Compute:

Press "A"__________.2571 inches (minimum wall thickness)

Optional: Compute design pressure for a given wall thickness.
Given:
wall thickness = .250
all other conditions from prior example

KEYSTROKES	OUTPUT
.250 "A"	.250
Press "B"	1,400 psig (compute design pressure)

STEP	KEY ENTRY	KEY CODE
001	f LBL A	31 25 11
	STO 1	33 01
	h F? 3	35 71 03
	h RTN	35 22
	RCL 2	34 02
	RCL 3	34 03
	x	71
	2	02
	÷	81
010	RCL 4	34 04
	÷	81
	RCL 5	34 05
	÷	81
	DSP 4	23 04
	f -x-	31 84
	STO 1	33 01
	h RTN	35 22
	f LBL B	31 25 12
	STO 2	33 02
020	h F? 3	35 71 03
	h RTN	35 22
	RCL 1	34 01
	2	02
	x	71
	RCL 4	34 04
	x	71
	RCL 5	34 05
	x	71
	RCL 3	34 03
030	÷	81
	DSP 4	23 04
	f -x-	31 84
	STO 2	33 02
	h RTN	35 22
	f LBL C	31 25 13
	STO 3	33 03
	h F? 3	35 71 03
	h RTN	35 22
	STO 3	33 03
040	h RTN	35 22
	f LBL D	31 25 14
	STO 4	33 04
	h F? 3	35 71 03
	h RTN	35 22
	STO 4	33 04
	h RTN	35 22
	f LBL E	31 25 15
	STO 5	33 05
	h F? 3	35 71 03
050	h RTN	35 22
	STO 5	33 05
	h RTN	35 22
	g LBLf a	32 25 11
	f CL REG	31 43
	h RTN	35 22

3.0 Preliminary Pipeline Sizing and BHP Computations

3.1 Introduction

The objectives of this program are to select a viable pipeline diameter (gas, liquid, two-phase flow) and compute the approximate BHP required to initialize a feasibility study.

3.2 Theoretical Formulae

(1) Liquids

$$v = \frac{.0119Q}{I.D.^2}$$

where:

v = velocity - ft./sec.
Q = flowrate - bpd
I.D. = internal diameter - inches

Economic Design Velocities: 5 - 8 ft./sec.
Hot Oil Line: 3 - 5 ft./sec.

$$BHP = \frac{(P)(B/H)}{2450 \text{ eff}}$$

where:

P = total discharge pressure
B/H = flowrate B/H
eff = pump efficiency - ratio
BHP = Brake horsepower

(2) Gases

$$\bar{u} = \frac{4QbV}{\pi \, d^2 \, V_b}$$

where:

Q_b = flowrate - scf/sec.
d = I.D. of pipe - feet
V = specific volume @ operating conditions
V_b = specific volume @ standard conditions
u = mean velocity - ft./sec.

$$V = \frac{R_o \, T_o \, z_o}{(29)(G)(P_o)} \qquad V_b = \frac{R \, T_b \, z_b}{(29)(G)(P_b)}$$

where:

R = 10.73
T = 460 + °F (flowing temperature)

z = compressibility factor
G = specific gravity of gas

Subscripts:
o = operating conditions
b = base conditions

Economic Design Velocities:

Natural Gas Gathering System: 35 ft./sec. - 45 ft./sec.
Natural Gas Transmission System: 12 - 35 ft./sec.

(3) Gas Line Fill:

$$\frac{P_1V_1}{T_1z_1} = \frac{P_2V_2}{T_2z_2} = 1.955d^2\ L\ P_m\ (1/z)^{.5}\quad ft^3/mi.$$

OR: $(5.28)(I.D.)^2\ (.372)(P)$ (in cubic ft./mi.)

(4) Pressure Drop: 10 - 15 PSI/mi.
Economic range for Gathering and Transmission Systems.

(5) $$BHP/MMcfd = \frac{R}{R + RJ}\left[\frac{5.16 + 124\ \log R}{.97 - .03\ R}\right]$$

where:
R = compression ratio
J = supercompressibility factor - .022/100 psia suction pressure

(6) Two-Phase Flow

Design Velocity: 10 - 15 ft./sec. gas superficial velocity

LP-F-003 - Preliminary Pipeline Sizing & BHP Computation
(Liquid, Gas, Two-Phase Flow)

Input:

	(1)	liquid throughput - bpd
OPTIONAL	(2)	pump efficiency - n
OPTIONAL	(3)	compression ratio - n
OPTIONAL	(4)	total pressure drop (liquids) - PSI (PSI/mi. x mi.)
	(5)	average pressure (gas) - psig
	(6)	gas throughput - mcfd
	(7)	gas specific gravity - n
	(8)	gas flowing temperature - ^{o}F
	(9)	gas base temperature - ^{o}F
	(10)	gas base pressure - psia
	(11)	gas compressibility factor - n
OPTIONAL	(12)	gas suction pressure for compressor - psia

Input/Computes:

(1) internal diameter of liquid line/gas line
(2) velocity of flow - liquid line/gas line

Compute:

(1) BHP - liquid pipeline
(2) BHP/MMCFD - gas pipeline
(3) gas density @ standard conditions
(4) gas density @ operating conditions

STEP	INSTRUCTIONS	KEYS	OUTPUT UNITS
1.	load card LP-F-003, side 1 & 2		
2.	input liquid thruput flowrate	A	bpd
3.	input liquid pump efficiency	A	n
4.	input compressor compression ratio (P_d/P_s = n)	A	n
5.	input total liquid line pressure drop (PSI/mi. x mi.)	A	PSI
6.	input average line pressure of gas	A	psig
7.	input gas flowrate	A	mcfd
8.	input gas specific gravity	A	n
9.	input gas flowing temperature	A	^{o}F
10.	input gas base temperature	A	^{o}F
11.	input gas base pressure	A	psia
12.	input compressibility factor	A	n
13.	input suction pressure for compressor	A	psia
	Liquids Pipeline		
14.	input/compute velocity	C	ft./sec.

STEP	INSTRUCTIONS	KEYS	OUTPUT UNITS
15.	input/compute internal diameter	B	inches
16.	compute total BHP	fb	BHP
	Gas Pipeline		
17.	* input/compute internal diameter	D	ft./inches
18.	input/compute velocity	E	ft./sec.

*Note: When inputing I.D. use inches.
When computing I.D., multiply by 12 to get inches.

STEP	INSTRUCTIONS	KEYS	OUTPUT UNITS
19.	compute BHP/MMcfd for gas	fd	BHP/MMcfd
20.	compute gas density at standard conditions	fc	ft.^{3}/lb.
		h 1/x	lb./ft.^{3}
21.	compute gas density at operating conditions	fe	ft.^{3}/lb.
		h 1/x	lb./ft.^{3}
22.	to clear and reset program	fa	

Example #1: LP-F-003 - Liquids Pipeline Preliminary Sizing and BHP requirements Computation

Given Conditions:
gasoline flowrate: 20,000 bpd
pump efficiency: .77
total pressure drop: (3.79 PSI/mi.) x (39 mi.): 147.81 PSI

Execution: Load card LP-F-003, side 1 & 2

Store parameters in this order:

KEYSTROKES	OUTPUT
20,000 "A"	20,000.00
.77 "A"	.77
0.00 "A"	0.00 (necessary input)
147.81 "A"	147.81

To Compute Internal Diameter:

6.0 "C"__________6.0 (good design velocity, 6 ft./sec.)

Press "B"__________6.3 inches (internal diameter)

Optional: Compute flow velocity:

6.3 "B"__________6.3 inches (input I.D.)

Press "C"__________6.0 ft./sec. (compute flow velocity)

Optional:

KEYSTROKES	OUTPUT
"fb"________________	65.29 BHP
"fa"________________	reset program

Example #2: LP-F-003

Given: flowrate = 52,910 bpd, LPG

Compute an approximate I.D. @ 6 ft./sec. flow velocity.

Execution: Load card LP-F-003

KEYSTROKES	OUTPUT
52,910 "A"__________	52,910.00
6 "C"_______________	6.00 (input flow velocity)
Press "B"___________	10.24 inches (compute I.D.)

Example #3: LP-F-003 - Gas Pipeline Sizing and Compressor BHP Requirements

Given Conditions:

gas specific gravity: .62
gas flowrate: 47,000 mcfd
average pressure: 288 psia
flowing temperature: 40°F
base temperature: 60°F
compressibility factor: computed below
suction pressure: 250 psia
discharge pressure: 1250 psia
base pressure: 14.73 psia

Execution: Load card GP-F-001 (gas compressibility) side 1 & 2

KEYSTROKES	OUTPUT
.62 "A"____________	669.84 (P_c)
288 "R/S"__________	366 (T_c)
500 "R/S"__________	.0849 (constant)

Guess z, approximately .6
.6 "E"______________ .94 (compute compressibility)

Load card LP-F-003, side 1 & 2

Store parameters in this order:

KEYSTROKES	OUTPUT
0.00 "A"	0.00 (necessary zero input)
0.00 "A"	0.00 (necessary zero input)
5 "A"	5.00 (P_d/P_s compression ratio)
0.00 "A"	0.00 (necessary zero input)
288 "A"	288.00
47,000 "A"	47,000.00
.62 "A"	.62
40 "A"	40.00
60 "A"	60.00
14.73 "A"	14.73
.94 "A"	.94
250 "A"	250.00

To Compute:

35 "E"	35.00 (input velocity) (35 ft./sec., good design velocity)
Press "D"	.96 feet (11.48") (compute I.D.)
"fd"	106.15 (compute BHP/MMcfd)
"106.15 ENTER↑ 47 x"	4,989.15 BHP
"fc"	21.07 ($1/\rho$)
"h 1/x"	.05 lb./ft.^{3} @ standard conditions
"fe"	.97 ft.^{3}/lb. @ operating conditions
"h 1/x"	1.03 lb./ft.^{3} (density at operating conditions)
"fa"	reset program

Example #4: LP-F-003 - Two Phase Flow Preliminary Line Sizing and BHP Requirements Computation

Given Conditions:

condensate flowrate: 3,700 bpd
gas flowrate: 96,000 mcfd
S.G. condensate: .61
S.G. gas: .62
pump efficiency: .77
compression ratio: 5
total pressure required: 347 PSI (liquids - well to plant)
average gas pressure: 288 psia
flowing temperature: 40°F
base temperature: 60°F
base pressure: 14.73 psia

suction pressure: 250 psia (gas) → compressor located at
discharge pressure: 1250 psia (gas) plant inlet

Execution: Compute preliminary two phase flow internal diameter for the given parameters.

Load card GP-F-001, side 1 & 2

Compute Compressibility:

KEYSTROKES	OUTPUT
.62 "A"	669.84 (P_c)
288 "R/S"	366 (T_c)
500 "R/S"	.0849 (a constant)

Guess a Factor: .7

.7 "E"	.94 (Newton approximation for compressibility factor)

Store parameters in this order:

3700 "A"	3,700.00
.77 "A"	.77
5 "A"	5.00
347 "A"	347.00
288 "A"	288.00
96,000 "A"	96,000.00
.62 "A"	.62
40 "A"	40.00
60 "A"	60.00
14.73 "A"	14.73
.94 "A"	.94 (compressibility)
250 "A"	250.00
6 "C"	6.00 (6 ft./sec. liquid velocity)

To Compute:

Press "B"	2.71 inches (I.D. Condensate)
35 "E"	35.00 ft./sec. (gas velocity)
Press "D"	1.37 feet (I.D. gas)
1.37 x 12	16.40 inches

Summation of approximate internal diameter = 19.11 inches. Good approximation of actual two phase flow internal diameter.

"fd"	106.15 BHP/MMcfd
"106.15 x 96"	10,190.61 BHP total

KEYSTROKES	OUTPUT
"fe"	.97 ft.^{3}/lb.
"h 1/x"	1.03 lb./ft.^{3} (density at operating conditions)
"fb"	28.36 (BHP for condensate alone)

STEP	KEY ENTRY	KEY CODE
001	f LBL A	31 25 11
	STO (i)	33 24
	f GSB 1	31 22 01
	h RTN	35 22
	f LBL 1	31 25 01
	f ISZ	31 34
	R/S	84
	h ST I	35 33
	GTO 1	22 01
010	h RTN	35 22
	f LBL B	31 25 12
	STO A	33 11
	h F? 3	35 71 03
	h RTN	35 22
	RCL 0	34 00
	.	83
	0	00
	1	01
	1	01
020	9	09
	x	71
	RCL B	34 12
	÷	81
	f $\sqrt{x}$	31 54
	STO A	33 11
	h RTN	35 22
	f LBL C	31 25 13
	STO B	33 12
	h F? 3	35 71 03
030	h RTN	35 22
	RCL 0	34 00
	.	83
	0	00
	1	01
	1	01
	9	09
	x	71
	RCL A	34 11
	g x^2	32 54
040	÷	81
	STO B	33 12
	h RTN	35 22
	f LBL D	31 25 14
	1	01
	2	02
	÷	81
	STO C	33 13
	h F? 3	35 71 03
	h RTN	35 22
050	RCL 5	34 05
	1	01
	0	00
	0	00
	0	00
	x	71
	2	02

STEP	KEY ENTRY	KEY CODE
	4	04
	÷	81
	6	06
060	0	00
	÷	81
	6	06
	0	00
	÷	81
	4	04
	x	71
	h π	35 73
	÷	81
	RCL D	34 14
070	÷	81
	g GSB c	32 22 13
	÷	81
	g GSB e	32 22 15
	x	71
	f $\sqrt{x}$	31 54
	STO C	33 13
	h RTN	35 22
	f LBL E	31 25 15
	STO D	33 14
080	h F? 3	35 71 03
	h RTN	35 22
	RCL 5	34 05
	1	01
	0	00
	0	00
	0	00
	x	71
	2	02
	4	04
090	÷	81
	6	06
	0	00
	÷	81
	6	06
	0	00
	÷	81
	4	04
	x	71
	h π	35 73
100	÷	81
	RCL C	34 13
	g x^2	32 54
	÷	81
	g GSB c	32 22 13
	÷	81
	g GSB e	32 22 15
	x	71
	STO D	33 14
	h RTN	35 22
110	g LBLf c	32 25 13
	1	01
	0	00

STEP	KEY ENTRY	KEY CODE
	.	83
	7	07
	3	03
	RCL 8	34 08
	4	04
	6	06
	0	00
120	+	61
	x	71
	2	02
	9	09
	÷	81
	RCL 6	34 06
	÷	81
	RCL 9	34 09
	÷	81
	h RTN	35 22
130	g LBLf e	32 25 15
	1	01
	0	00
	.	83
	7	07
	3	03
	RCL 7	34 07
	4	04
	6	06
	0	00
140	+	61
	x	71
	2	02
	9	09
	÷	81
	RCL 6	34 06
	÷	81
	RCL 4	34 04
	÷	81
	f P⪌S	31 42
150	RCL 0	34 00
	x	71
	f P⪌S	31 42
	h RTN	35 22
	g LBLf a	32 25 11
	f CL REG	31 43
	f P⪌S	31 42
	f CL REG	31 43
	h RTN	35 22
	g LBLf b	32 25 12
160	RCL 3	34 03
	RCL 0	34 00
	2	02
	4	04
	÷	81
	x	71
	2	02
	4	04
	5	05
	0	00
170	÷	81
	RCL 1	34 01
	÷	81
	h RTN	35 22
	g LBLf d	32 25 14
	RCL 2	34 02
	f LOG	31 53
	1	01
	2	02
	4	04
180	x	71
	5	05
	.	83
	1	01
	6	06
	+	61
	STO (i)	33 24
	RCL 2	34 02
	.	83
	0	00
190	3	03
	x	71
	CHS	42
	.	83
	9	09
	7	07
	+	61
	RCL (i)	34 24
	h x⪌y	35 52
	÷	81
200	STO (i)	33 24
	RCL 2	34 02
	f P⪌S	31 42
	RCL 1	34 01
	f P⪌S	31 42
	1	01
	0	00
	0	00
	÷	81
	.	83
210	0	00
	2	02
	2	02
	x	71
	x	71
	RCL 2	34 02
	+	61
	RCL 2	34 02
	h x⪌y	35 52
	÷	81
220	RCL (i)	34 24
	x	71
	h RTN	35 22

References

3-1 Wilson & McAdams, "Contribution No. 19", Chemical Engineering Department, M.I.T.

3-2 AGA, "Steady Flow in Gas Pipelines", Institute of Gas Technology, Report No. 10, July 1965.

3-3 NGPSA, Engineering Data Book, 1972.

4.0 Gas Compressibility

4.1 Introduction

This computation is utilized frequently in gas computations.

4.2 Theoretical Formulae - Benedict-Webb-Rubin Equation

$$Z = 1 + (A_1 + A_2/T_r + A_3/T_r^3)\rho_r + (A_4 + A_5/T_r)\rho_r^2 + A_5A_6\rho_r^5/T_r + A_7\rho_r^2/T_r^3\ (1 + A_8\rho_r^2)\ \exp\ (-A_8\rho_r^2)$$

Where:

ρ_r = 0.27 $P_r/(zT_r)$ - psuedo-reduced density
T_r = psuedo-reduced temperature
P_r = psuedo-reduced pressure
z = compressibility factor
A_1 = 0.3151
A_2 = -1.0467
A_3 = -0.5783
A_4 = 0.5353
A_5 = -0.6123
A_6 = -0.1049
A_7 = 0.6816
A_8 = 0.6845

GP-F-001 - Gas compressibility Computation

Input:

(1) gas specific gravity - n
(2) gas average pressure - psia
(3) flowing gas temperature - ^{o}R (460 + ^{o}F)

Computes:

(1) compressibility of gas at average pressure and temperature conditions.
(2) critical pressure - psia
(3) critical temperature - ^{o}R
(4) reduced pressure - n
(5) reduced temperature - n

STEP	INSTRUCTIONS	KEYS	OUTPUT UNITS
1.	load card GP-F-001, side 1 & 2		
2.	input specific gravity of gas	A	psia (P_c)
3.	input average gas pipeline pressure (psia)	R/S	^{o}R (T_c)
4.	input flowing gas temperature (°F)	R/S	constant
5.	guess a compressibility input for Newtonian approximation	E	n
6.	recall for reduced pressure	RCL B	P_r
7.	recall for reduced temperature	RCL C	T_v
8.	for new case	f CL REG	

Example: GP-F-001 - Compressibility Computation

Given Conditions:

gas specific gravity: .62
flowing gas temperature: 40°F
inlet pressure: 1415 psia
outlet pressure: 1295 psia
average pressure: 1363 psia

Execution: Load card GP-F-001

KEYSTROKES	OUTPUT
.62 "A"	669.84 (P_c)
1363 "R/S"	366.00 (T_c)
500 "R/S"	.402 (a constant)

KEYSTROKES	OUTPUT
.6 "E"	.736 (guess input for Newton approximation of compressibility)
"RCL B"	2.03 (P_r)
"RCL C"	1.366 (T_r)
"f CL REG"	new case - resets program

STEP	KEY ENTRY	KEY CODE
001	f LBL E	31 25 15
	f FIX	31 23
	GTO B	22 12
	g LBLf a	32 25 11
	RCL 2	34 02
	GTO 0	22 00
	f LBL 6	31 25 06
	RCL 0	34 00
	f GSB 9	31 22 09
010	STO 2	33 02
	f LBL 0	31 25 00
	RCL 1	34 01
	RCL 0	34 00
	STO 1	33 01
	-	51
	RCL 4	34 04
	RCL 2	34 02
	STO 4	33 04
	-	51
020	÷	81
	x	71
	STO -0	33 51 00
	RCL 0	34 00
	h PAUSE	35 72
	÷	81
	f RND	31 24
	f x≠0	31 61
	GTO 6	22 06
	RCL 0	34 00
030	h RTN	35 22
	f LBL B	31 25 12
	.	83
	0	00
	1	01
	f %	31 82
	STO 3	33 03
	2	02
	÷	81
	-	51
040	STO 1	33 01
	STO 0	33 00
	f GSB 9	31 22 09
	STO 4	33 04
	RCL 1	34 01
	RCL 3	34 03
	+	61
	STO 0	33 00
	f GSB 9	31 22 09
	STO 2	33 02
050	RCL 4	34 04
	-	51
	RCL 3	34 03
	÷	81
	GTO f a	22 31 11
	f LBL 9	31 25 09
		83

STEP	KEY ENTRY	KEY CODE
	5	05
	7	07
	8	08
060	CHS	42
	RCL 7	34 07
	3	03
	h y^x	35 63
	÷	81
	1	01
	.	83
	0	00
	4	04
	6	06
070	7	07
	CHS	42
	RCL 7	34 07
	÷	81
	.	83
	3	03
	1	01
	5	05
	+	61
	+	61
080	RCL 8	34 08
	RCL 0	34 00
	÷	81
	x	71
	1	01
	+	61
	h ST I	35 33
	.	83
	6	06
	1	01
090	2	02
	3	03
	CHS	42
	RCL 7	34 07
	÷	81
	.	83
	5	05
	3	03
	5	05
	+	61
100	RCL 8	34 08
	RCL 0	34 00
	÷	81
	g x^2	32 54
	x	71
	h RC I	35 34
	+	61
	h ST I	35 33
	RCL 8	34 08
	RCL 0	34 00
110	÷	81
	5	05
	h y^x	35 63

STEP	KEY ENTRY	KEY CODE
	.	83
	1	01
	0	00
	4	04
	CHS	42
	x	71
	.	83
120	6	06
	1	01
	2	02
	3	03
	CHS	42
	x	71
	RCL 7	34 07
	÷	81
	h RC I	35 34
	+	61
130	h ST I	35 33
	RCL 8	34 08
	RCL 0	34 00
	÷	81
	g x^2	32 54
	RCL 7	34 07
	3	03
	h y^x	35 63
	÷	81
	.	83
140	6	06
	8	08
	1	01
	5	05
	7	07
	x	71
	STO 5	33 05
	RCL 8	34 08
	RCL 0	34 00
	÷	81
150	g x^2	32 54
	.	83
	6	06
	8	08
	4	04
	5	05
	x	71
	1	01
	+	61
	RCL 5	34 05
160	x	71
	STO 5	33 05
	RCL 8	34 08
	RCL 0	34 00
	÷	81
	g x^2	32 54
	.	83
	6	06
	8	08

STEP	KEY ENTRY	KEY CODE
	4	04
170	5	05
	CHS	42
	x	71
	g e^x	32 52
	RCL 5	34 05
	x	71
	h RC I	35 34
	+	61
	RCL 0	34 00
	CHS	42
180	+	61
	h RTN	35 22
	f LBL A	31 25 11
	STO A	33 11
	4	04
	3	03
	x	71
	CHS	42
	6	06
	9	09
190	6	06
	.	83
	5	05
	+	61
	R/S	84
	h x≷y	35 52
	÷	81
	STO 6	33 06
	STO B	33 12
	RCL A	34 11
200	3	03
	0	00
	0	00
	x	71
	1	01
	8	08
	0	00
	+	61
	R/S	84
	h x≷y	35 52
210	÷	81
	STO C	33 13
	STO 7	33 07
	RCL 6	34 06
	h x≷y	35 52
	÷	81
	.	83
	2	02
	7	07
	x	71
220	STO 8	33 08
	h RTN	35 22

References

4-1 Dranchuk, P.M., et al, "Computer Calculation of Natural Gas Compressibility Factors Using the Standing and Katz Correlation", Petroleum Society of CIM

4-2 NGPSA Engineering Data Book, 1972.

4-3 Benedict, M., et al, "An Empirical Equation for Thermodynamic Properties of Light Hydrocarbons and Their Mixtures", J. Chem. Phys. Vol. 8, pp. 334-45, 1940.

5.0 Slurry Feasibility

5.1 Preliminary Pipe Size Selection and Critical Velocity

5.2 Introduction

This program executes a preliminary computation of pipe size and fluid requirements at a velocity of 5.5 ft./sec. and solids concentrations as stated in section #4 of the theoretical formulae. Additionally, it can execute a preliminary pressure drop computation. These computations apply to horizontal flow only.

5.3 Theoretical Formulae

(1) Homogeneous Flow Pressure Drop

$$i_m = i_w\ (1 + C_v(s - 1))$$

(2) Heterogeneous Flow Pressure Drop

$$I_m = i_w \left[1 + 81 \times C_v \left[\frac{gD(s - 1)}{V^2} \times \frac{1}{\sqrt{C_D}}\right]^{1.5}\right]$$

$$V\text{critical}_1 = F_L\ (2gD(s - 1))^{.5} \quad \text{- (Durand)}$$

$$V\text{critical}_2 = \text{two-phase vehicle} \quad \text{- (Wasp et al)}$$

$$V\text{transition} = (1800\ gDV_s)^{1/3} \quad \text{- (Newitt)}$$

WHERE:

C_D = drag coefficient
C_v = concentration of solids by volume
g = acceleration due to gravity
i_m = hydraulic gradient ft. water/length of pipe (slurry)
i_w = hydraulic gradient ft. water/length of pipe (water)
s = solid density (ratio)
$V\text{critical}_1$ = deposition velocity - ft./sec.
$V\text{critical}_2$ = deposition velocity - ft./sec.
Vtransition = homogeneous/heterogeneous boundary velocity - ft./sec.
V = slurry flow velocity - ft./sec.
F_L = Durand factor

(3) Conversion factors

1 U.S. gallon = 8.345 lbs. (water)
1 cubic foot = 7.4805 U.S. gallons
1 pound of water = .01602 ft.^{3}

$$\underset{\text{(ft./sec.)}}{\text{velocity}} = \frac{.4085\text{x(U.S. gallons/minute)}}{D^2\ \text{(inches)}}$$

$$\underset{\text{(U.S. gpm)}}{\text{flowrate}} = \frac{D^2\ \text{(inches)}\ V\text{(ft./sec.)}}{.4085}$$

$$1\ ft.^3/min. = \frac{D^2(inches)\ V(ft./sec.)}{.0546}$$

$$= .13367 \times U.S.\ gpm$$

$$\text{slurry horsepower} = \frac{(gpm)(head)(s.g.)(8.33)}{33{,}000}$$

$$BHP = \frac{\text{slurry horsepower}}{\text{pump efficiency}}$$

$$kwh/1000\ gallons/HR = \frac{(head)(s.g.)(.00315)}{(\text{pump eff.})(\text{motor eff.})}$$

(4) Minimum Particle Sizes

	Solid s.g.	Maximum Particle Size	Average Slurry Concentration % Solids by Weight
Coal	1.4	8 mesh	50
Limestone	2.7	48 mesh	70
Phosphate	3.2	48 mesh	65
Copper Concentrate	4.3	65 mesh	55
Iron Concentrate	4.9	100 mesh	60

SP-F-001 - Preliminary Pipe Size Selection & Critical Velocity Computation

Input:

(1) slurry flowrate - $ft.^3/min.$
(2) internal diameter - inches
(3) F_L factor - n
(4) diameter of particle - inches
(5) settling velocity of particle - ft./sec.
(6) density of solid - n (ratio)
(7) type of particle, ie.
  (a) coal = 2
  (b) iron ore = 3
  (c) limestone = 4
  (d) copper ore = 5
(8) slurry flowrate - TONS/yr.

Computes:

(1) transport velocity - ft./sec.
(2) critical velocity - ft./sec. (Durand)
(3) critical velocity - ft./sec. (Wasp et.al.)
(4) transition velocity - ft./sec. (Newitt)
  (boundary between homogeneous & heterogeneous flow)
(5) approximate I.D. for given flowrate
(6) approximate acre-ft./yr. water requirements
(7) approximate gallons/yr. water requirements

STEP	INSTRUCTIONS	KEYS	OUTPUT UNITS
1.	load card SP-F-001, side 1 & 2		
2.	input volumetric flowrate of product	E	MM TONS/yr.
3.	input appropriate number for product to be transported, ie.	ENTER↑	n
	2 = coal		
	3 = iron ore		
	4 = limestone		
	5 = copper ore		
	Compute:	D	
	(1) approximate I.D. @5.5 ft./sec. flow velocity		inches
	(2) water volume required		acre-ft./yr.
	(3) water volume required		U.S. gal./yr.
4.	input volume of slurry	A	$ft.^3/min.$
5.	input I.D. of pipeline	A	inches
6.	input F_L factor	A	n
7.	input diameter of particle	A	inches
8.	input particle settling velocity	A	ft./sec.
9.	input solid density	A	n

STEP	INSTRUCTIONS	KEYS	OUTPUT UNITS
10.	Compute:	B	
	(1) transport velocity		ft./sec.
	(2) critical velocity (Durand equation)		ft./sec.
	(3) critical velocity (Wasp et.al.)		ft./sec.
	(4) transition velocity (Newitt et.al.)		ft./sec.
11.	Optional:		
	To clear for new case	fa	
12.	Option to Vary Parameters:		
	(1) input new particle diameter	fb	inches
	(2) input new settling velocity for this particle	fc	ft./sec.
	(3) input new internal diameter	fd	inches
	(4) to compute new velocity parameters	B	

Example: SP-F-001 - Approximate I.D. & Critical Velocity Computation

Given Conditions:

Volume of slurry: 1.3 MM TONS/yr. of coal to be transported

Compute approximate I.D. and water requirements:

Execution: Load card SP-F-001, side 1 & 2

	KEYSTROKES	OUTPUT
	1.3 "E"	1.3
To Compute:	"2 ENTER↑"	2.0
	"D"	"9.5" inches
		"650" acre-ft./yr. water
		"3.79×10^6" U.S. gal./yr. of water

Compute transport, critical and transition velocities for this I.D.:

Given Conditions:

volume of slurry: 142.5 $ft.^3$/min.
internal diameter: 9.5 inches
F_L factor: 1.0
diameter of particle: .0331 inches
particle settling velocity: .003 ft./sec.
solid density: 1.4 (coal)

Store parameters in this order:

KEYSTROKES	OUTPUT
142.5 "A"	142.5000
9.5 "A"	9.5000

KEYSTROKES	OUTPUT
1.0 "A"	1.0000
.0331 "A"	.0331
.003 "A"	.0030
1.4 "A"	1.4000

To Compute:

Press "B"	4.82 ft./sec. (transport velocity)
Press "C"	"4.52" ft./sec. (critical velocity - Durand)
	"1.8" ft./sec. (critical velocity - Wasp et.al.)
	"5.15" ft./sec. (transition velocity - Newitt et.al.)

STEP	KEY ENTRY	KEY CODE
001	f LBL A	31 25 11
	STO (i)	33 24
	f GSB 1	31 22 01
	h RTN	35 22
	f LBL 1	31 25 01
	f ISZ	31 34
	R/S	84
	h ST I	35 33
	GTO 1	22 01
010	h RTN	35 22
	f LBL B	31 25 12
	RCL 0	34 00
	3	03
	.	83
	0	00
	5	05
	x	71
	RCL 1	34 01
	g x^2	32 54
020	÷	81
	STO A	33 11
	h RTN	35 22
	f LBL C	31 25 13
	RCL 5	34 05
	1	01
	CHS	42
	+	61
	6	06
	4	04
030	.	83
	4	04
	x	71
	RCL 1	34 01
	1	01
	2	02
	÷	81
	x	71
	f $\sqrt{x}$	31 54
	h PAUSE	35 72
040	STO B	33 12
	RCL 3	34 03
	RCL 1	34 01
	÷	81
	6	06
	h 1/x	35 62
	h y^x	35 63
	x	71
	STO C	33 13
	h PAUSE	35 72
050	RCL 1	34 01
	1	01
	2	02
	÷	81
	RCL 4	34 04
	x	71
	3	03
	2	02
	.	83
	2	02
060	x	71
	1	01
	8	08
	0	00
	0	00
	x	71
	.	83
	3	03
	3	03
	3	03
070	h y^x	35 63
	STO D	33 14
	h RTN	35 22
	f LBL D	31 25 14
	h ST I	35 33
	f GSB (i)	31 22 24
	f -x-	31 84
	f GSB 6	31 22 06
	f -x-	31 84
	5	05
080	8	08
	2	02
	3	03
	.	83
	5	05
	3	03
	x	71
	f -x-	31 84
	0	00
	h ST I	35 33
090	h RTN	35 22
	f LBL 2	31 25 02
	RCL E	34 15
	.	83
	4	04
	9	09
	h y^x	35 63
	8	08
	.	83
	3	03
100	8	08
	x	71
	h RTN	35 22
	f LBL 3	31 25 03
	RCL E	34 15
	.	83
	5	05
	h y^x	35 63
	5	05
	.	83
110	7	07
	8	08
	x	71

STEP	KEY ENTRY	KEY CODE
	h RTN	35 22
	f LBL 4	31 25 04
	RCL E	34 15
	.	83
	6	06
	3	03
	h y^x	35 63
120	4	04
	.	83
	6	06
	5	05
	x	71
	h RTN	35 22
	f LBL 5	31 25 05
	RCL E	34 15
	.	83
	6	06
130	h y^x	35 63
	7	07
	.	83
	1	01
	6	06
	x	71
	h RTN	35 22
	f LBL 6	31 25 06
	RCL E	34 15
	1	01
140	h y^x	35 63
	5	05
	0	00
	0	00
	x	71
	h RTN	35 22
	f LBL E	31 25 15
	STO E	33 15
	h RTN	35 22
	g LBLf a	32 25 11
150	f CL REG	31 43
	f P≷S	31 42
	f CL REG	31 43
	h RTN	35 22
	g LBLf b	32 25 12
	STO 3	33 03
	h RTN	35 22
	g LBLf c	32 25 13
	STO 4	33 04
	h RTN	35 22
160	g LBLf d	32 25 14
	STO 1	33 01
	h RTN	35 22

SP-F-002 - Pressure Drop and BHP Computation

Input:

(1) flow velocity - slurry
(2) internal diameter - inches
(3) solid density - ratio
(4) solid volume percent - ratio
(5) drag coefficient of particle - n
(6) length of pipeline - miles
(7) pump efficiency - n (decimal)
(8) pressure drop per mile - PSI/mi.

Computes:

(1) pressure drop for water
  (a) Reynolds No. modified
  (b) pressure drop for water - PSI/ft.
  (c) pressure drop for water - ft. water/ft.
(2) pressure drop for homogeneous flow - PSI/mi.
(3) pressure drop for heterogeneous flow - PSI/mi.
(4) BHP requirements

STEP	INSTRUCTIONS	KEYS	OUTPUT UNITS
1.	load card SP-F-002, side 1 & 2		
2.	input flow velocity	A	ft./sec.
3.	input internal diameter	A	inches
4.	input solid density	A	ratio
5.	input solid volume percent	A	(decimal)
6.	input drag coefficient of particle (heterogeneous - water)	A	n
7.	input length of pipeline	A	miles
8.	input pump efficiency	A	n
9.	Compute water pressure drop computation:	B	
	(1) Reynolds no. for water modified		n
	(2) pressure drop (water)		PSI/ft.
	(3) pressure drop (water)		ft./ft.
10.	Water pressure drop Computation:	fb	
	(1) Reynolds No. modified		n
	(2) pressure drop (water)		PSI/ft.
	(3) pressure drop (water)		PSI/mi.
11.	Compute homogeneous pressure drop	C	PSI/mi.
12.	Compute heterogeneous pressure drop	D	PSI/mi.
13.	input pressure drop per mile for horsepower computation	ENTER↑, fe	PSI/mi.
14.	Compute total horsepower requirements:	E	
	(1) flowrate		gpm
	(2) total pressure drop		PSI
	(3) total BHP		BHP

STEP	INSTRUCTIONS	KEYS	OUTPUT UNITS
15.	Optional:		
	(1) input a new pressure drop per mile for BHP computation	fe	PSI/mi.
	(2) Compute:	E	
	(a) flowrate		gpm
	(b) total pressure drop		PSI
	(c) total BHP		BHP
16.	Optional: To clear and reset program for new case:	fa	

Example: SP-F-002 - Pressure Drop & BHP Computations

Given Conditions:

flow velocity: 6.84 ft./sec.
internal diameter: 8 inches
solid density: 1.4
volume % solid: .40
drag coefficient: 13.7 (heterogeneous case)
length of pipeline: 50 miles
pump efficiency: .77

Execution: Load card SP-F-002, side 1 & 2

Store parameters in this order:

KEYSTROKES	OUTPUT
6.84 "A"	6.84
8.00 "A"	8.00
1.4 "A"	1.40
.4 "A"	.40
13.7 "A"	13.70
50.0 "A"	50.00
.77 "A"	.77

To Compute Water Pressure Drop:

Press "B"__________ "54.72" (Reynolds No. modified)
__________ ".00642" (PSI/ft. - water)
__________ ".01483" (ft./ft. - water)

Press "fb"__________ "54.72" (Reynolds No. modified)
__________ ".00642" (PSI/ft. - water)
__________ "33.90" (PSI/mi. - water)

Press "C"__________ 39.32 PSI/mi. (homogeneous pressure drop)

KEYSTROKES	OUTPUT
Press "D"	46.03 PSI/mi. (heterogeneous pressure drop)
46.03 ENTER	46.03
"fe"	46.03
Press "E"	"1072.94" gpm
	"2301.5" PSI
	"1870.0" H.P.
39.32 ENTER	39.32
"fe"	39.32
Press "E"	"1072.94" gpm
	"1966.0" PSI
	"1597.4" BHP
"fa"	clears and resets for new case

STEP	KEY ENTRY	KEY CODE
001	f LBL A	31 25 11
	STO (i)	33 24
	f GSB 1	31 22 01
	h RTN	35 22
	f LBL 1	31 25 01
	f ISZ	31 34
	R/S	84
	h ST I	35 33
	GTO 1	22 01
010	h RTN	35 22
	f LBL B	31 25 12
	RCL 0	34 00
	g x^2	32 54
	.	83
	3	03
	2	02
	3	03
	x	71
	RCL 1	34 01
020	÷	81
	STO A	33 11
	RCL 1	34 01
	RCL 0	34 00
	x	71
	h PAUSE	35 72
	.	83
	4	04
	g $x \le y$	32 71
	GTO 8	22 08
030	GTO 7	22 07
	h RTN	35 22
	f LBL 8	31 25 08
	h $x \gtrless y$	35 52
	h 1/x	35 62
	.	83
	3	03
	5	05
	5	05
	h y^x	35 63
040	.	83
	0	00
	0	00
	6	06
	6	06
	2	02
	x	71
	.	83
	0	00
	0	00
050	1	01
	8	08
	+	61
	RCL A	34 11
	x	71
	h PAUSE	35 72
		83

STEP	KEY ENTRY	KEY CODE
	4	04
	3	03
	3	03
060	÷	81
	STO B	33 12
	h RTN	35 22
	f LBL 7	31 25 07
	h $x \gtrless y$	35 52
	h 1/x	35 62
	.	83
	0	00
	0	00
	2	02
070	0	00
	7	07
	x	71
	RCL A	34 11
	x	71
	h PAUSE	35 72
	.	83
	4	04
	3	03
	3	03
080	÷	81
	STO B	33 12
	h RTN	35 22
	f LBL C	31 25 13
	RCL 2	34 02
	1	01
	CHS	42
	+	61
	RCL 3	34 03
	x	71
090	1	01
	+	61
	RCL B	34 12
	x	71
	.	83
	4	04
	3	03
	3	03
	x	71
	5	05
100	2	02
	8	08
	0	00
	x	71
	STO C	33 13
	h RTN	35 22
	f LBL D	31 25 14
	RCL 2	34 02
	1	01
	CHS	42
110	+	61
	3	03
	2	02

STEP	KEY ENTRY	KEY CODE
	.	83
	2	02
	x	71
	RCL 1	34 01
	1	01
	2	02
	÷	81
120	x	71
	RCL 0	34 00
	g x^2	32 54
	÷	81
	RCL 4	34 04
	h 1/x	35 62
	f $\sqrt{x}$	31 54
	x	71
	1	01
	.	83
130	5	05
	h y^x	35 63
	RCL 3	34 03
	x	71
	8	08
	1	01
	x	71
	1	01
	+	61
	RCL B	34 12
140	x	71
	.	83
	4	04
	3	03
	3	03
	x	71
	5	05
	2	02
	8	08
	0	00
150	x	71
	STO D	33 14
	h RTN	35 22
	f LBL E	31 25 15
	RCL 0	34 00
	RCL 1	34 01
	g x^2	32 54
	x	71
	.	83
	0	00
160	1	01
	1	01
	9	09
	÷	81
	4	04
	2	02
	x	71
	2	02
	4	04

STEP	KEY ENTRY	KEY CODE
	÷	81
170	6	06
	0	00
	÷	81
	STO 9	33 09
	h PAUSE	35 72
	RCL 7	34 07
	RCL 5	34 05
	x	71
	h PAUSE	35 72
	RCL 9	34 09
180	x	71
	1	01
	7	07
	1	01
	5	05
	÷	81
	RCL 6	34 06
	÷	81
	h RTN	35 22
	g LBLf e	32 25 15
190	STO 7	33 07
	h RTN	35 22
	g LBLf a	32 25 11
	f CL REG	31 43
	h RTN	35 22
	g LBLf b	32 25 12
	f GSB B	31 22 12
	.	83
	4	04
	3	03
200	3	03
	x	71
	5	05
	2	02
	8	08
	0	00
	x	71
	h RTN	35 22

References

1-1 Colorado School of Mines Research Foundation, "Transportation of Solids in Steel Pipelines", 1963.

1-2 Wasp et al, "Solid-Liquid Flow, Slurry Pipeline Transportation," 1977.

1-3 Daniels, B.E., "Some Practical Considerations for the Solids Pipeline Engineer", Proceedings of the 2nd International Technical Conference on Slurry Transportation, March 2-4, 1977.

6.0 Economics

6.1 Introduction

To implement a fully turn-key approach to the three phases of pipeline operations, that is feasibility, design and construction, it is necessary to have an economic evaluation during each phase of the system under consideration. This section then is supplementary and is to be used as a guide only.

The economic programs can also serve as a self-contained feasibility package.

ECONOMICS COMPUTATIONS PROCEDURES

1. Compute Investment in $

   (a) Pipelines @ $/I.D. inch/mi. ______________

   (b) Pump stations @ $/BHP ______________

   (c) Compressor Stations @ $/BHP ______________

   (d) Tankage @ $/BBL ______________

   (e) Line Fill @ $/BBL and/or $/MCF, ie., part of working capital ______________

   TOTAL INVESTMENT ==============

2. Compute construction period in years ______________

3. Compute salvage value of facilities, $ ______________

4. Compute debt fraction ______________

5. Compute % annual interest on debt fraction ______________

6. Operating Expenses per Year

   A. Fixed Operating Expenses/yr.

      (1) supervisory, technical & clerical staff @ $ man/yr. ______________

      (2) operating personnel @ $ man/yr. ______________

      (3) taxes & insurance @ 1.5% ______________

      (4) pipeline & station maintenance @ 1.5% ______________

      (5) miscellaneous @ .5% ______________

      TOTAL FIXED EXPENSES ==============

   B. Variable Operating Expenses/yr.

      (1) Fuel & Utilities

         (a) Electric - centrifugal, driver & pump

      Given Rate Schedule

Demand Charge

First _____ KW or less @ $ __________

Next _____ KW @ $ /KW __________

Excess KW @ $ /KW __________

Energy Charge

First _______ KWH @ __________ ¢/KWH

Next _______ KWH @ __________ ¢/KWH

Next _______ KWH @ __________ ¢/KWH

Next _______ KWH @ __________ ¢/KWH

Excess _______ KWH @ __________ ¢/KWH

Compute Average HP monthly requirements:

Demand KW = HP x 1.08 x 1746 KW/HP

Energy KWH = HP x .746 x 30 days/mo. x 23.5 H/D
= __________ KWH/month

(b) Turbine-Centrifugal Pump

Given Rate Schedule

Natural Gas @ ____________ ¢/MCF

Average HP requirements/mo. ____________

HP_{avg} x 23.5 H/D x 30 days/mo. ____________ HP-hrs.

BTUs required HP-hrs.
= HP-hrs. x ___________ BTU/HP-hrs. x $\frac{1}{.20}$

Monthly Cost:
MCF x $/MCF = ___________ $/mo.

Average HP requirements:

$$= \frac{\text{B/H x PSI x .0004083}}{.80 \times .97}$$

(c) Compressor Station Fuel Requirements:

BHP x 240 = MCF fuel requirements/day

1 HP-hr. = 2,547 BTU

(d) Cone roof vapor losses: _____BBL @ _____$/BBL
Floating roof vapor losses: _____BBL @ _____$/BBL

(e) gas/product handling losses:
_________% of throughput

(f) Fuels:

Natural Gas = 1,000,000 BTU/MCF

Diesel = 150,000 BTU/gal.

(g) BTU Computation:
8,000 BTU/BHP-hr.

7. Royalties ______________

8. BTU Penalty ______________

9. <u>Revenues</u>

Thruput __________ BBLS/DAY @ $__________/BBL

Thruput __________ MCFD @ $__________/MCF

10. <u>Compute Rate of Return</u>

Gross Revenue
(minus) - Gross Expenses
" - Depreciation
" - Interest
" - CIOH (Capitalized Interest & Overhead)
" - Taxes

= NET INCOME AFTER TAXES

11. Compute Internal Rate of Return/Discount Cash Flow - Net Present Value Payout - years.

PL-E-001 - Cash Throw per Year Computation

Input:

(1) flowrate - bpd/mmcfd
(2) revenue - ¢/BBL & ¢/mcf
(3) investment - $
(4) fixed expenses - $/yr.
(5) pump efficiency - n
(6) average discharge pressure/pump - psig
(7) average cost of electricity - ¢/KWH
(8) average cost of gas - ¢/mcf
(9) average cost of product (liquids) - $/BBL
(10) no. of pump stations - n
(11) compressor station suction pressure - psig
(12) compressor station discharge pressure - psig

Computes:

(1) BHP requirements for the entire system - BHP
(2) total revenues for year one - ± M$/yr.
(3) fuel costs - M$/yr.
(4) fixed costs - M$/yr.
(5) cash throw - M$/yr.

STEP	INSTRUCTIONS	KEYS	OUTPUT UNITS
1.	load card PL-E-001, side 1 & 2		
2.	input volume flowrate	A	bpd/mmcfd
3.	input revenues	A	¢/BBL - ¢/mcf
4.	input total investment	A	$
5.	input fixed expenses	A	$/yr.
6.	input pump efficiency	A	n
7.	input average pump station discharge pressure	A	psig
8.	input average cost of electricity	A	¢/KWH
9.	input average cost of fuel gas	A	¢/mcf
10.	input average cost of transported product	A	$/BBL
11.	input no. of pump/compressor stations	A	n
12.	input average compressor station suction pressure	A	psig
13.	input average compressor station discharge pressure	A	psig
14.	Option: to input any miscellaneous expenses before taxes, interest or depreciation	STO E	M$/yr.
15.	Compute: Cash Throw (liquids only)	B	
	(1) BHP for total stations		BHP
	(2) cash throw per year		M$

STEP	INSTRUCTIONS	KEYS	OUTPUT UNITS
16.	Optional: To Compute-(liquids)		
	(1) total revenues	RCL A	M$/yr.
	(2) fuel costs	RCL B	M$/yr.
	(3) fixed expenses	RCL D	M$/yr.
	(4) miscellaneous expenses	RCL E	M$/yr.
17.	Compute Cash Throw: (gas only)	C	
	(1) compression ratio		n
	(2) BHP requirements for all stations		BHP/mmcfd
	(3) total BHP requirements		BHP
	(4) total fuel consumption		mcfd
	(5) total cash throw		M$/yr.
18.	Optional: To Compute-(gas)		
	(1) total revenue	RCL A	M$/yr.
	(2) total fuel costs	RCL C	M$/yr.
	(3) total fixed expenses	RCL D	M$/yr.
	(4) total miscellansous expenses	RCL E	M$/yr.
19.	Option: To Compute Pump Station BHP Requirements & Power cost/yr. (option)		
	(1) input flowrate	A	bpd
	(2) input zero	A	0.00
	(3) input zero	A	0.00
	(4) input zero	A	0.00
	(5) input pump efficiency	A	n
	(6) input average pump discharge pressure	A	psig
	(7) input power cost or input zero (optional)	A	¢/KWH - 0.00
	(8) input zero	A	0.00
	(9) input zero	A	0.00
	(10) input no. of stations	A	n
	(11) input zero	A	0.00
	(12) input zero	A	0.00
	(13) Compute:	D	
	(1) BHP total requirements		BHP
	(2) Optional: power cost if input in step 7.		¢/KWH
20.	Option: To Compute Compressor Station BHP Requirements & Fuel Cost/yr. (option)		
	(1) input flowrate	A	mmcfd
	(2) input zero	A	0.00
	(3) input zero	A	0.00
	(4) input zero	A	0.00
	(5) input zero	A	0.00
	(6) input zero	A	0.00

STEP	INSTRUCTIONS	KEYS	OUTPUT UNITS
	(7) input fuel cost or input zero (optional)	A	M$/yr. or 0.00
	(8) input zero	A	0.00
	(9) input no. of stations	A	n
	(10) input suction pressure	A	psig
	(11) input discharge pressure	A	psig
	(12) Compute:	E	
	(1) compression ratio		n (ratio)
	(2) brake horsepower/MMcfd		n
	(3) brake horsepower		n
	(4) fuel consumption		mcfd
	(5) Optional: fuel costs if input in step 7.		M$/yr.
21.	To reset program for new case	fa	

Note: All input parameters which are not applicable must be inputed with a zero value, ie. for a gas system, step 6 "pump efficiency" input must be zero.

Example: PL-E-001 - Cash Throw Computation for a Gas Gathering System

Given Conditions:

line length: 23 miles
O.D.: 12 inches
flowrate: 100 MMcfd
revenues: 12¢/mcf
investment: $40,000,000
fixed expenses: $1,200,000/yr.
fuel gas cost: 26¢/mcf
no. of stations: 1
compressor suction pressure: 185.3 psig
compressor discharge pressure: 985.3 psig

Execution: Load card PL-E-001, side 1 & 2

Store parameters in this order:

KEYSTROKES	OUTPUT
100 "A"	100.00
12 "A"	12.00
40,000,000 "A"	40,000,000.00
1,200,000 "A"	1,200,000.00
0.00 "A"	0.00
0.00 "A"	0.00
0.00 "A"	0.00
26 "A"	26.00
0.00 "A"	0.00

KEYSTROKES	OUTPUT
1.00 "A"	1.00
185.3 "A"	185.30
985.3 "A"	985.30

To Compute:

Press "C"	"5.00" (compression ratio)
	"107.27" (BHP/MMcfd)
	"10,727" (total BHP)
	"2,574.5" mcfd (fuel consumption)
	"2,935.68" M$/yr. (cash throw)
"RCL A"	4,380 M$/yr. (revenues)
"RCL C"	244.32 M$/yr. (fuel costs)
"RCL D"	1,200 M$/yr. (fixed costs)
"RCL E"	0.00 M$/yr. (miscellaneous costs)

Option: To compute fuel costs only-

Press "E"	"5.00" (compression ratio)
	"107.27" BHP/MMcfd
	"10,727" (total BHP)
	"2,574.5" mcfd (fuel consumption)
	"244.32" M$ (fuel cost)

Example: PL-E-001 - Cash Throw Computation for a Liquids Pipeline Gathering System

Given Conditions:

line length: 12 miles
O.D.: 6 inches
flowrate: 5000 bpd
revenues: 5¢/BBL
investment: $800,000
fixed expenses: $24,000
pump efficiency: .77
average discharge pressure: 800 PSI
average cost of electricity: 1¢/KWH
average cost of product: $6/BBL
no. of pump stations: 1

Execution: Load card PL-E-001, side 1 & 2

Store parameters in this order:

KEYSTROKES	OUTPUT
5000 "A"	5,000.00
5 "A"	5.00
800,000 "A"	800,000.00
24,000 "A"	24,000.00

KEYSTROKES	OUTPUT
.77 "A"	.77
800 "A"	800.00
1 "A"	1.00
0.00 "A"	0.00
6 "A"	6.00
1 "A"	1.00
0.00 "A"	0.00
0.00 "A"	0.00

To Compute:

Press "B"	"88.38" BHP (total BHP)
	"61.23" M$/yr. (cash throw)
"RCL A"	91.25 M$/yr. (revenues)
"RCL B"	6.02 M$/yr. (fuel costs)
"RCL D"	24.00 M$/yr. (fixed expenses)
"RCL E"	0.00 M$/yr. (miscellaneous expenses)
"fa"	to reset program for new case

STEP	KEY ENTRY	KEY CODE
001	f LBL A	31 25 11
	STO (i)	33 24
	f GSB 1	31 22 01
	h RTN	35 22
	f LBL 1	31 25 01
	f ISZ	31 34
	R/S	84
	h ST I	35 33
	GTO 1	22 01
010	h RTN	35 22
	f LBL 2	31 25 02
	RCL 0	34 00
	3	03
	6	06
	5	05
	x	71
	RCL 1	34 01
	x	71
	1	01
020	0	00
	0	00
	÷	81
	3	03
	CHS	42
	g 10^x	32 53
	x	71
	STO A	33 11
	h RTN	35 22
	f LBL 3	31 25 03
030	RCL 3	34 03
	1	01
	0	00
	0	00
	0	00
	÷	81
	STO D	33 14
	h RTN	35 22
	f LBL D	31 25 14
	RCL 0	34 00
040	2	02
	4	04
	÷	81
	RCL 5	34 05
	x	71
	.	83
	0	00
	0	00
	0	00
	4	04
050	0	00
	8	08
	3	03
	x	71
	RCL 4	34 04
	÷	81
	h PAUSE	35 72

STEP	KEY ENTRY	KEY CODE
	1	01
	.	83
	0	00
060	8	08
	x	71
	.	83
	7	07
	4	04
	6	06
	x	71
	3	03
	0	00
	x	71
070	2	02
	3	03
	.	83
	5	05
	x	71
	1	01
	2	02
	x	71
	RCL 6	34 06
	x	71
080	1	01
	0	00
	0	00
	÷	81
	1	01
	0	00
	0	00
	0	00
	÷	81
	RCL 9	34 09
090	x	71
	STO B	33 12
	h RTN	35 22
	f LBL E	31 25 15
	f $P\gtrless S$	31 42
	RCL 1	34 01
	1	01
	4	04
	.	83
	7	07
100	+	61
	RCL 0	34 00
	1	01
	4	04
	.	83
	7	07
	+	61
	÷	81
	h PAUSE	35 72
	STO C	33 13
110	f LOG	31 53
	1	01
	2	02

STEP	KEY ENTRY	KEY CODE
	4	04
	x	71
	5	05
	.	83
	1	01
	6	06
	+	61
120	STO D	33 14
	RCL C	34 13
	.	83
	0	00
	3	03
	x	71
	CHS	42
	.	83
	9	09
	7	07
130	+	61
	RCL D	34 14
	h x≷y	35 52
	÷	81
	STO D	33 14
	RCL 0	34 00
	1	01
	4	04
	.	83
	7	07
140	+	61
	1	01
	0	00
	0	00
	÷	81
	.	83
	0	00
	2	02
	2	02
	x	71
150	RCL C	34 13
	x	71
	RCL C	34 13
	+	61
	RCL C	34 13
	h x≷y	35 52
	÷	81
	RCL D	34 14
	x	71
	h PAUSE	35 72
160	f P≷S	31 42
	RCL 0	34 00
	f P≷S	31 42
	x	71
	h PAUSE	35 72
	.	83
	2	02
	4	04
	0	00

STEP	KEY ENTRY	KEY CODE
	x	71
170	h PAUSE	35 72
	f P≷S	31 42
	RCL 7	34 07
	x	71
	1	01
	0	00
	0	00
	÷	81
	RCL 9	34 09
	x	71
180	3	03
	6	06
	5	05
	x	71
	1	01
	0	00
	0	00
	0	00
	÷	81
	STO C	33 13
190	h RTN	35 22
	f LBL B	31 25 12
	f GSB 2	31 22 02
	f GSB 3	31 22 03
	f GSB D	31 22 14
	RCL A	34 11
	RCL D	34 14
	-	51
	RCL B	34 12
	-	51
200	RCL E	34 15
	-	51
	h RTN	35 22
	f LBL C	31 25 13
	f GSB 2	31 22 02
	1	01
	0	00
	0	00
	0	00
	x	71
210	STO A	33 11
	f GSB E	31 22 15
	f GSB 3	31 22 03
	RCL A	34 11
	RCL C	34 13
	-	51
	RCL D	34 14
	-	51
	RCL E	34 15
	-	51
220	h RTN	35 22
	g LBLf a	32 25 11
	f CL REG	31 43
	f P≷S	31 42
	h RTN	35 22

PL-E-002 - Break Even Analysis Computation

Input/Compute:

(1) fixed costs - \$
(2) cost/BBL, cost/mcf - \$/BBL. \$/mcf
(3) variable expenses - \$/BBL, \$/mcf
(4) flowrate - BBL/yr., mcf/yr.
(5) cash throw per year - \$/yr.

STEP	INSTRUCTIONS	KEYS	OUTPUT UNITS
1.	load card PL-E-002, side 1		
2.	Input four of the following in any order:		
	(1) fixed costs	A	\$/yr.
	(2) revenues per BBL or mcf	B	\$/BBL, \$/mcf
	(3) variable costs per BBL or mcf	C	\$/BBL, \$/mcf
	(4) number of BBLS or mcf	D	n
	(5) gross profit	E	\$
3.	Compute the remaining variable by pressing "A, B, C, D or E" as required.		
4.	Compute: degree of operating leverage	fa	n (ratio)

Example: PL-E-002 - Break Even Analysis Computation

Given Conditions: An oil gathering system has the following parameters-

fixed expenses per year: \$24,000
flowrate per year: 500 bpd x 365 = 1,825,000 BBLS
variable expenses: \$.004/BBL (fuel costs)
cash throw required: \$200,000/yr.
total investment: \$820,000

Execution: Load card PL-E-002, side 1

Store parameters in any order:

KEYSTROKES	OUTPUT
24,000 "A"	24,000.000
.004 "C"	.004
1,825,000 "D"	1,825,000.000
250,000 "E"	250,000.000

To Compute:

Press "B"__________\$.15/BBL (required revenues)

STEP	KEY ENTRY	KEY CODE
001	f LBL A	31 25 11
	STO A	33 11
	h F? 3	35 71 03
	h RTN	35 22
	RCL B	34 12
	RCL C	34 13
	-	51
	RCL D	34 14
	x	71
010	RCL E	34 15
	-	51
	STO A	33 11
	h RTN	35 22
	f LBL B	31 25 12
	STO B	33 12
	h F? 3	35 71 03
	h RTN	35 22
	RCL A	34 11
	RCL E	34 15
020	+	61
	RCL D	34 14
	÷	81
	RCL C	34 13
	+	61
	STO B	33 12
	h RTN	35 22
	f LBL C	31 25 13
	STO C	33 13
	h F? 3	35 71 03
030	h RTN	35 22
	RCL B	34 12
	RCL A	34 11
	RCL E	34 15
	+	61
	RCL D	34 14
	÷	81
	-	51
	STO C	33 13
	h RTN	35 22
040	f LBL D	31 25 14
	STO D	33 14
	h F? 3	35 71 03
	h RTN	35 22
	RCL A	34 11
	RCL E	34 15
	+	61
	RCL B	34 12
	RCL C	34 13
	-	51
050	÷	81
	STO D	33 14
	h RTN	35 22
	f LBL E	31 25 15
	STO E	34 15
	h F? 3	35 71 03
	h RTN	35 22
	RCL B	34 12
	RCL C	34 13
	-	51
060	RCL D	34 14
	x	71
	RCL A	34 11
	-	51
	STO E	33 15
	h RTN	35 22
	g LBLf a	32 25 11
	RCL B	34 12
	RCL C	34 13
	-	51
070	RCL D	34 14
	x	71
	h ST I	35 33
	h RC I	35 34
	RCL A	34 11
	-	51
	÷	81
	h RTN	35 22

PL-E-003 - Amortization Schedule Computation

Input:

(1) interest rate - %
(2) constant payment to principal - $
(3) initial loan amount - $

Computes:

(1) payment to interest per year - $
(2) total payment per year - $
(3) remaining balance - $
(4) total interest - $

STEP	INSTRUCTIONS	KEYS	OUTPUT UNITS
1.	load card PL-E-003, side 1		
2.	input first period of the desired schedule	A	n
3.	input interest rate	B	%
4.	input constant payment to principle	C	$
5.	input initial loan amount	D	$
6.	Compute:		
	(1) payment to interest per year	E	$
	(2) total payment per year	R/S	$
	(3) remaining balance per year	R/S	$
	(4) total interest	R/S	$
7.	To begin next period of schedule	R/S	n

Example: PL-E-003 - Amortization Schedule Computation

A 20 year, 8% loan for $100,000 is being amortized by annual payments to principle of $5,000 plus interest on the remaining balance. Generate a 2 year amortization schedule.

Given Conditions:

interest rate: 8%
constant payment to principle: $5,000
initial loan amount: $100,000
amortization period: 20 years

Execution: Load card PL-E-003, side 1

Store parameters in this order:

KEYSTROKES	OUTPUT
1 "A"	1.00
8 "B"	8.00

KEYSTROKES	OUTPUT
5000 "C"	5,000.00
100,000 "D"	100,000.00

To Compute:

Press "E"	\$8,000.00	(1st year's payment to interest)
"R/S"	\$13,000.00	(total 1st payment)
"R/S"	\$95,000.00	(remaining balance)
"R/S"	\$8,000.00	(total interest paid to date)
"R/S"	2.00	(now starting 2nd period)
"R/S"	\$7,600.00	(2nd year's payment to interest)
"R/S"	\$12,600.00	(total 2nd payment)
"R/S"	\$90,000.00	(remaining balance)
"R/S"	\$15,600.00	(total interest paid to date)

STEP	KEY ENTRY	KEY CODE
001	f LBL A	31 25 11
	h CF 1	35 61 01
	STO 0	33 00
	f GSB 9	31 22 09
	h RTN	35 22
	f LBL B	31 25 12
	EEX	43
	2	02
	÷	81
010	STO B	33 12
	h LST x	35 82
	x	71
	h RTN	35 22
	f LBL C	31 25 13
	STO C	33 13
	h RTN	35 22
	f LBL D	31 25 14
	STO D	33 14
	h RTN	35 22
020	f LBL E	31 25 15
	RCL D	34 14
	RCL C	34 13
	RCL 0	34 00
	x	71
	-	51
	STO E	33 15
	RCL C	34 13
	+	61
	RCL B	34 12
030	x	71
	STO 9	33 09
	1	01
	STO +0	33 61 00
	RCL 9	34 09
	f GSB 9	31 22 09
	RCL C	34 13
	+	61
	f GSB 9	31 22 09
	RCL E	34 15
040	f GSB 9	31 22 09
	f LBL 0	31 25 00
	2	02
	RCL 0	34 00
	-	51
	RCL C	34 13
	x	71
	RCL D	34 14
	÷	81
	2	02
050	+	61
	2	02
	÷	81
	RCL 0	34 00
	1	01
	-	51
	x	71

STEP	KEY ENTRY	KEY CODE
	RCL B	34 12
	x	71
	RCL D	34 14
060	x	71
	h F? 1	35 71 01
	h RTN	35 22
	f GSB 9	31 22 09
	RCL D	34 14
	RCL C	34 13
	÷	81
	RCL 0	34 00
	g x>y	32 81
	h RTN	35 22
070	f -x-	35 84
	f GSB 9	31 22 09
	GTO E	22 15
	g LBLf a	32 25 11
	h SF 1	35 51 01
	STO 0	33 00
	h x≷y	35 52
	STO 8	33 08
	1	01
	STO +0	33 61 00
080	f GSB 0	31 22 00
	h ST I	35 33
	RCL 8	34 08
	STO 0	33 00
	f GSB 0	31 22 00
	h RC I	35 34
	h x≷y	35 52
	-	51
	f GSB 9	31 22 09
	h RTN	35 22
090	g LBLf e	32 25 15
	h F? 0	35 71 00
	GTO 1	22 01
	h SF 0	35 51 00
	1	01
	h RTN	35 22
	f LBL 1	31 25 01
	0	00
	h CF 0	35 61 00
	h RTN	35 22
100	f LBL 9	31 25 09
	h F? 0	35 71 00
	GTO 2	22 02
	R/S	84
	h RTN	35 22
	f LBL 2	31 25 02
	f -x-	31 84
	h RTN	35 22

PL-E-004 - Depreciation Schedules Computation

Input:

(1) starting book value - $
(2) salvage value - $
(3) life of the asset - years
(4) year for which depreciation is to be calculated - n

Computes:

(1) straight line depreciation schedule
(2) sum-of-the-years-digits depreciation schedule
(3) declining balance depreciation schedule
(4) annual depreciation amount - $
(5) remaining depreciable amount - $
(6) remaining book value - $
(7) total depreciation to date - $

STEP	INSTRUCTIONS	KEYS	OUTPUT UNITS
1.	load card PL-E-004, side 1 & 2		
2.	input starting book value	STO A	$
3.	input salvage value	STO B	$
4.	input life of the asset	STO C	years
5.	For depreciation schedules, input: year for which depreciation is to be calculated	STO D	n
6.	Compute: Straight Line Depreciation Schedule-		
	(1) year	fa	n
	(2) annual depreciation amount	R/S	$
	(3) remaining depreciable amount	R/S	$
	(4) remaining book value	R/S	$
	(5) total depreciation to date	R/S	$
	(6) year + 1	R/S	n
Note:	Key R/S to keep calculating step #6(2-6)		
7.	Optional: For new case go to steps #2-5 and change appropriate inputs.		
8.	Compute: Sum-of-the-Years-Digits Depreciation Schedule-		
	(1) year	fb	n
	(2) annual depreciation amount	R/S	$
	(3) remaining depreciable amount	R/S	$
	(4) remaining book value	R/S	$
	(5) total depreciation to date	R/S	$
	(6) year + 1	R/S	n
Note:	Key R/S to keep calculating step #8(2-6)		
9.	Optional: For new case go to steps #2-5 and change appropriate inputs.		

STEP	INSTRUCTIONS	KEYS	OUTPUT UNITS
10.	input appropriate factor for Declining Balance Schedule	STO E	n
11.	Compute: Declining Balance Schedule		
	(1) year	fc	n
	(2) annual depreciation amount	R/S	$
	(3) remaining depreciable amount	R/S	$
	(4) remaining book value	R/S	$
	(5) total depreciation to date	R/S	$
	(6) year + 1	R/S	n
	Key "R/S" to keep calculating steps #11(2-6)		
12.	Optional: For new case go to steps #2-5 and change appropriate inputs.		
13.	To Compute crossover point:		
	(1) input declining balance factor	STO E	n
	(2) compute last year to use declining balance method	fd	n
	(3) compute remaining life	R/S	n
	(4) compute remaining book value	R/S	$

Example: PL-E-004 - Depreciation Schedules Computation

For a starting book value of $375,000, a salvage value of $30,000, and an expected life of 40 years, generate the 1st year's depreciation schedule using each of the common methods. Assume a declining balance factor of 1.5. Then jump ahead to the 15th year and generate the data for that year.

Given Conditions:

starting book value: $375,000
salvage value: $30,000
expected life of asset: 40 years
declining balance factor: 1.5
year for which depreciation is to be calculated: 1st & 15th

Execution: Load card PL-E-004, side 1 & 2

Store parameters in this order:

KEYSTROKES	OUTPUT
375,000 "STO A"	375,000.00
30,000 "STO B"	30,000.00
40 "STO C"	40.00
1 "STO D"	1.00

To Compute: Straight Line Depreciation Schedule-

KEYSTROKES	OUTPUT
"fa"	1.00 (1st year)
"R/S"	$8,625.00 (1st year's depreciation)
"R/S"	$336,375.00 (remaining depreciable value)
"R/S"	$366,375.00 (remaining book value)
"R/S"	$8,625.00 (total depreciation to date)

To Compute 15th year:

15 "STO D, fa"	15.00 (15th year)
"R/S"	$8,625.00 (15th year's depreciation)
"R/S"	$215,625.00 (remaining depreciable value)
"R/S"	$245,625.00 (remaining book value)
"R/S"	$129,375.00 (total depreciation after 15 years)

To Compute: Sum of the Years Digits Depreciation Schedule-

1 "STO D, fb"	1.00 (1st year)
"R/S"	$16,829.27 (1st year's depreciation)
"R/S"	$328,170.73 (remaining depreciable value)
"R/S"	$358,170.73 (remaining book value)
"R/S"	$16,829.27 (total depreciation to date)

To Compute 15th year:

15 "STO D, fb"	15.00 (15th year)
"R/S"	$10,939.02 (15th year's depreciation)
"R/S"	$136,737.80 (remaining depreciable value)
"R/S"	$166,737.80 (remaining book value)
"R/S"	$208,262.20 (total depreciation 1st through 15th year)

To Compute: Declining Balance Depreciation Schedule-

KEYSTROKES	OUTPUT
1 "STO D"	
1.5 "STO E, fc"	1.00 (1st year)
"R/S"	$14,062.50 (1st year's depreciation)
"R/S"	$330,937.50 (remaining depreciable value)
"R/S"	$360,937.50 (remaining book value)
"R/S"	$14,062.50 (total depreciation to date)

To Compute 15th year:

15 "STO D, fc"	15.00 (15th year)
"R/S"	$8,235.18 (15th year's depreciation)
"R/S"	$181,369.51 (remaining depreciable value)
"R/S"	$211,369.51 (remaining book value)
"R/S"	$163,630.49 (total depreciation 1st through 15th year)

STEP	KEY ENTRY	KEY CODE
001	g LBLf a	32 25 11
	h F? 0	35 71 00
	h SPACE	35 84
	RCL D	34 14
	f GSB 9	31 22 09
	RCL A	34 11
	RCL B	34 12
	-	51
	RCL C	34 13
010	÷	81
	h ST I	35 33
	f GSB 9	31 22 09
	RCL C	34 13
	RCL D	34 14
	-	51
	h RC I	35 34
	x	71
	f GSB 9	31 22 09
	RCL B	34 12
020	+	61
	f GSB 9	31 22 09
	h RC I	35 34
	RCL D	34 14
	x	71
	f GSB 9	31 22 09
	1	01
	f GSB D	31 22 14
	RCL C	34 13
	RCL D	34 14
030	g x≤y	32 71
	GTO f a	22 31 11
	h RTN	35 22
	g LBLf b	32 25 12
	h F? 0	35 71 00
	h SPACE	35 84
	RCL D	34 14
	f GSB 9	31 22 09
	RCL A	34 11
	RCL B	34 12
040	-	51
	STO 8	33 08
	RCL C	34 13
	f GSB 2	31 22 02
	STO 7	33 07
	RCL C	34 13
	1	01
	+	61
	RCL D	34 14
	-	51
050	RCL 7	34 07
	÷	81
	RCL 8	34 08
	x	71
	f GSB 9	31 22 09
	RCL C	34 13
	RCL D	34 14
	-	51
	f x<0	31 71
	GTO 3	22 03
060	f GSB 2	31 22 02
	RCL 7	34 07
	÷	81
	STO 4	33 04
	RCL 8	34 08
	x	71
	f LBL 3	31 25 03
	STO 6	33 06
	f GSB 9	31 22 09
	RCL B	34 12
070	+	61
	f GSB 9	31 22 09
	1	01
	RCL 4	34 04
	-	51
	RCL 8	34 08
	x	71
	f GSB 9	31 22 09
	1	01
	f GSB D	31 22 14
080	RCL C	34 13
	RCL D	34 14
	g x≤y	32 71
	GTO f b	22 31 12
	h RTN	35 22
	f LBL 2	31 25 02
	ENTER ↑	41
	g FRAC	32 83
	ENTER ↑	41
	+	61
090	h x≷y	35 52
	f INT	31 83
	+	61
	h LST x	35 82
	1	01
	+	61
	x	71
	2	02
	÷	81
	h RTN	35 22
100	g LBLf c	32 25 13
	h F? 0	35 71 00
	h SPACE	35 84
	RCL D	34 14
	f GSB 9	31 22 09
	f GSB 4	31 22 04
	RCL D	34 14
	1	01
	-	51
	h y^x	35 63
110	RCL A	34 11
	x	71
	RCL 8	34 08

STEP	KEY ENTRY	KEY CODE
	x	71
	h ST I	35 33
	f GSB 9	31 22 09
	1	01
	RCL 7	34 07
	RCL D	34 14
	h y^x	35 63
120	-	51
	RCL A	34 11
	x	71
	STO 9	33 09
	RCL A	34 11
	RCL B	34 12
	-	51
	RCL 9	34 09
	-	51
	f GSB 9	31 22 09
130	RCL B	34 12
	+	61
	f GSB 9	31 22 09
	RCL 9	34 09
	f GSB 9	31 22 09
	1	01
	f GSB D	31 22 14
	RCL C	34 13
	RCL D	34 14
	g x≤y	32 71
140	GTO f c	22 31 13
	h RTN	35 22
	f LBL D	31 25 14
	RCL D	34 14
	+	61
	STO D	33 14
	h RTN	35 22
	f LBL 4	31 25 04
	1	01
	RCL E	34 15
150	RCL C	34 13
	÷	81
	STO 8	33 08
	-	51
	STO 7	33 07
	h RTN	35 22
	g LBLf e	32 25 15
	h F? 0	35 71 00
	GTO 0	22 00
	h SF 0	35 51 00
160	1	01
	h RTN	35 22
	f LBL 0	31 25 00
	0	00
	h CF 0	35 61 00
	h RTN	35 22
	f LBL 9	31 25 09
	h F? 0	35 71 00
	GTO 1	22 01

STEP	KEY ENTRY	KEY CODE
	R/S	84
170	h RTN	35 22
	f LBL 1	31 25 01
	f -x-	31 84
	h RTN	35 22
	g LBLf d	32 25 14
	0	00
	STO D	33 14
	f GSB 4	31 22 04
	f LBL 8	31 25 08
	RCL 7	34 07
180	1	01
	f GSB D	31 22 14
	1	01
	-	51
	h y^x	35 63
	RCL A	34 11
	x	71
	RCL 8	34 08
	x	71
	RCL 7	34 07
190	RCL D	34 14
	1	01
	-	51
	h y^x	35 63
	RCL A	34 11
	x	71
	RCL B	34 12
	-	51
	STO 9	33 09
	RCL C	34 13
200	1	01
	+	61
	RCL D	34 14
	-	51
	÷	81
	h x≷y	35 52
	g x>y	32 81
	GTO 8	22 08
	RCL D	34 14
	1	01
210	-	51
	f GSB 9	31 22 09
	RCL C	34 13
	h x≷y	35 52
	-	51
	f GSB 9	31 22 09
	RCL 9	34 09
	RCL B	34 12
	+	61
	GTO 9	22 09

PL-E-005 - Internal Rate of Return Computation

Input:

(1) initial investment - $
(2) largest cash flow if >22 cash flows exist - $
(3) each cash flow - $

Computes:

(1) periodic internal rate of return - %

STEP	INSTRUCTIONS	KEYS	OUTPUT UNITS
1.	load card PL-E-005, side 1 & 2		
2.	input initial investment	A	$
3.	input largest cash flow if >22 cash flows exist	B	$
4.	input all cash flows in sequence	C	$
5.	compute periodic internal rate of return	D	%

Example: PL-E-005 - Internal Rate of Return Computation

Pipeline equipment requiring a $250,000 equity investment and to be sold in 10 years is expected to generate the "after tax" cash flows shown below. What is the expected internal rate of return?

Given Conditions:

End of Year	Cash Flow
1	$46,423
2	40,710
3	36,638
4	34,097
5	32,485
6	23,199
7	21,612
8	20,037
9	18,460
10	311,406 (equipment sold)

Execution: Load card PL-E-005, side 1 & 2

Store parameters in this order:

KEYSTROKES	OUTPUT
250,000 "A"	250,000.00
46,423 "C"	46,423.00
40,710 "C"	40,710.00

KEYSTROKES	OUTPUT
36,638 "C"	36,638.00
34,097 "C"	34,097.00
32,485 "C"	32,485.00
23,199 "C"	23,199.00
21,612 "C"	21,612.00
20,037 "C"	20,037.00
18,460 "C"	18,460.00
311,406 "C"	311,406.00

To Compute:

Press "D" ______ 13.98% (annual internal rate of return)

Note:

The interest rate that equates the present value of all future cash flows with the original investment is known as the internal rate of return (IRR, also called discounted rate of return or yield). Given a non-zero initial investment and up to 44 positive cash flows, this program calculates the periodic IRR. If there are negative as well as positive cash flows, the program accepts up to 22 cash flows.

If more than 44 positive cash flows are entered, all cash flows over 44 will be ignored. There will be no indication, however, that more than 44 cash flows have been entered. Likewise, if more than 22 positive and negative cash flows are entered, erroneous results will occur.

Zero should be entered for periods with no cash flow.

When more than 22 cash flows are involved (all of which must be positive), the user is asked to enter the largest cash flow in step 3 because of the storage techniques being used. This value is then used to scale all other cash flows, and depending on these values, accuracy may be reduced. Consequently, the resulting periodic rate of return should be considered accurate to within $\pm$ .01% (.0001 decimal). This largest cash flow must be entered again in sequence in step 4. If a cash flow larger than the value entered for CF MAX is keyed in at step 4, erroneous results may occur.

STEP	KEY ENTRY	KEY CODE
001	f LBL A	31 25 11
	f CL REG	31 43
	f P≷S	31 42
	f CL REG	31 43
	STO E	33 15
	h CF 0	35 61 00
	h CF 1	35 61 01
	h RTN	35 22
	f LBL B	31 25 12
010	2	02
	x	71
	STO 0	33 00
	RCL E	34 15
	h x≷y	35 52
	÷	81
	STO E	33 15
	h LST x	35 82
	h SF 0	35 51 00
	2	02
020	÷	81
	h RTN	35 22
	f LBL C	31 25 13
	f ISZ	31 34
	h F? 0	35 71 00
	g GSB c	32 22 13
	STO +(i)	33 61 24
	h x≷y	35 52
	h RC I	35 34
	h F? 1	35 71 01
030	+	61
	h RTN	35 22
	g LBLf c	32 25 13
	2	02
	3	03
	h RC I	35 34
	g x≠y	32 61
	GTO 0	22 00
	1	01
	h ST I	35 33
040	+	61
	CL x	44
	EEX	43
	5	05
	STO ÷0	33 81 00
	h SF 1	35 51 01
	f LBL 0	31 25 00
	h R↓	35 53
	1	01
	-	51
050	h x≷y	35 52
	RCL 0	34 00
	÷	81
	h F? 1	35 71 01
	f INT	31 83
	h RTN	35 22
	g LBLf a	32 25 11

STEP	KEY ENTRY	KEY CODE
	h RC I	35 34
	1	01
	0	00
060	1	01
	x	71
	h ST I	35 33
	h RTN	35 22
	g LBLf e	32 25 15
	h F? 0	35 71 00
	GTO 0	22 00
	f INT	31 83
	EEX	43
	5	05
070	÷	81
	h RTN	35 22
	f LBL 0	31 25 00
	g FRAC	32 83
	h RTN	35 22
	f LBL D	31 25 14
	g GSB a	32 22 11
	h RC I	35 34
	EEX	43
	2	02
080	÷	81
	h ST I	35 33
	0	00
	h F? 1	35 71 01
	GTO 1	22 01
	f LBL 2	31 25 02
	RCL (i)	34 24
	+	61
	f DSZ	31 33
	GTO 2	22 02
090	GTO 0	22 00
	f LBL 1	31 25 01
	h CF 0	35 61 00
	f LBL 3	31 25 03
	RCL (i)	34 24
	g GSB e	32 22 15
	+	61
	f DSZ	31 33
	GTO 3	22 03
	h F? 0	35 71 00
100	GTO 0	22 00
	g GSB b	32 22 12
	h SF 0	35 51 00
	GTO 3	22 03
	f LBL 0	31 25 00
	RCL E	34 15
	-	51
	RCL E	34 15
	÷	81
	g GSB a	32 22 11
110	f INT	31 83
	h CF 0	35 61 00
	h F? 1	35 71 01

STEP	KEY ENTRY	KEY CODE
	g GSB d	32 22 14
	÷	81
	1	01
	+	61
	STO D	33 14
	f LBL 4	31 25 04
	h CF 0	35 61 00
120	0	00
	STO 0	33 00
	f LBL 5	31 25 05
	h RC I	35 34
	f INT	31 83
	h F? 1	35 71 01
	g GSB d	32 22 14
	RCL (i)	34 24
	h F? 1	35 71 01
	g GSB e	32 22 15
130	STO +0	33 61 00
	x	71
	+	61
	RCL D	34 14
	STO ÷0	33 81 00
	÷	81
	f DSZ	31 33
	GTO 5	22 05
	h F? 1	35 71 01
	GTO 0	22 00
140	f LBL 6	31 25 06
	RCL 0	34 00
	RCL E	34 15
	−	51
	h x≷y	35 52
	÷	81
	RCL D	34 14
	x	71
	RCL D	34 14
	h x≷y	35 52
150	+	61
	STO D	33 14
	h LST x	35 82
	h ABS	35 64
	EEX	43
	CHS	42
	5	05
	g x>y	32 81
	GTO 7	22 07
	g GSB a	32 22 11
160	GTO 4	22 04
	f LBL 0	31 25 00
	h F? 0	35 71 00
	GTO 6	22 06
	h SF 0	35 51 00
	g GSB b	32 22 12
	GTO 5	22 05
	g LBLf b	32 25 12
	2	02

STEP	KEY ENTRY	KEY CODE
	2	02
170	h RC I	35 34
	+	61
	h ST I	35 33
	CL x	44
	+	61
	h RTN	35 22
	g LBLf d	32 25 14
	2	02
	2	02
	h F? 0	35 71 00
180	CL x	44
	+	61
	h RTN	35 22
	f LBL 7	31 25 07
	RCL D	34 14
	1	01
	−	51
	STO D	33 14
	EEX	43
	2	02
190	x	71
	h RC I	35 34
	h LST x	35 82
	x	71
	h ST I	35 33
	h x≷y	35 52
	h RTN	35 22

PL-E-006 - Discounted cash Flow Analysis - Net Present Value Computation

Input:

(1) initial investment amount - \$
(2) periodic interest (discount) rate - %

Input/Computes:

(1) net present value per cash flow - \$

STEP	INSTRUCTIONS	KEYS	OUTPUT UNITS
1.	load card PL-E-006, side 1		
2.	input initial investment amount	A	\$
3.	input periodic interest rate	B	%
4.	input number of equal cash flows if greater than 1	C	n
5.	input cash flow amounts/compute net present value	D	\$
6.	Optional: Display total no. of cash flows entered so far	E	n
7.	For next cash flow(s) go to step #4		

Example: PL-E-006 - Discounted Cash Flow Analysis - Net Present Value Computation

An investor has an opportunity to purchase pipeline equipment for \$70,000. If the going rate of return on this type of investment is 13.75%, and the after-tax cash flows are forecast as follows, should the investor purchase the equipment?

Given Conditions:

Year	Cash Flow (\$)
1	14,000
2	11,000
3	10,000
4	10,000
5	10,000
6	9,100
7	9,000
8	9,000
9	4,500
10	71,000 (equipment sold in 10th year)

Execution: Load card PL-E-006, side 1

KEYSTROKES	OUTPUT
70,000 "A"	70,000.00 (input initial investment)
13.75 "B"	13.75 (rate of return - %)
14,000 "D"	-57,692.31 (net present value after 1 cash flow - $)
11,000 "D"	-49,190.92 (net present value after 2 cash flows - $)
3 "C"	3.00 (no. of equal cash flows)
10,000 "D"	-31,172.57 (net present value after 5 cash flows - $)
9100 "D"	-26,971.76 (net present value after 6 cash flows - $)
2 "C"	2.00 (no. of equal cash flows)
9000 "D"	-20,108.39 (net present value after 8 cash flows - $)
"E"	8.00 (checking that 8 cash flows have been entered so far)
4500 "D"	-18,696.99 (net present value after 9 cash flows - $)
71,000 "D"	879.93 (net present value after 10 cash flows - $)

Since the final net present value is positive, the investment meets the profit objectives.

STEP	KEY ENTRY	KEY CODE
001	f LBL A	31 25 11
	CHS	42
	STO A	33 11
	0	00
	STO 9	33 09
	1	01
	STO C	33 13
	RCL A	34 11
	CHS	42
010	f GSB 9	31 22 09
	h RTN	35 22
	f LBL B	31 25 12
	EEX	43
	2	02
	÷	81
	STO B	33 12
	h LST x	35 82
	x	71
	h RTN	35 22
020	f LBL C	31 25 13
	STO C	33 13
	h RTN	35 22
	f LBL D	31 25 14
	STO D	33 14
	1	01
	RCL B	34 12
	+	61
	RCL C	34 13
	STO +9	33 61 09
030	h y^x	35 63
	STO E	33 15
	RCL A	34 11
	x	71
	RCL E	34 15
	1	01
	-	51
	RCL B	34 12
	÷	81
	RCL D	34 14
040	x	71
	+	61
	STO A	33 11
	1	01
	RCL B	34 12
	+	61
	RCL 9	34 09
	h y^x	35 63
	÷	81
	1	01
050	STO C	33 13
	h R↓	35 53
	f GSB 9	31 22 09
	h RTN	35 22
	f LBL E	31 25 15
	RCL 9	34 09
	h RTN	35 22
	g LBLf e	32 25 15
	h F? 0	35 71 00
	GTO 1	22 01
060	h SF 0	35 51 00
	1	01
	h RTN	35 22
	f LBL 1	31 25 01
	0	00
	h CF 0	35 61 00
	h RTN	35 22
	f LBL 9	31 25 09
	h F? 0	35 71 00
	GTO 2	22 02
070	R/S	84
	h RTN	35 22
	f LBL 2	31 25 02
	f -x-	31 84

Example: Application of Economic Computations

A gas well is currently capable of producing 100 mmcfd of natural gas, specific gravity of .62.

It is proposed that an 18 inch pipeline be constructed to transport the gas from the wellsite to an existing central plant 7 miles away, where a booster compressor station will deliver the gas to the plant at 915 psia. The following parameters apply:

inlet pressure: 543 psia (from wellsite)
outlet pressure: 185.3 psia (delivery to compressors)
compressor station discharge: 985.3 psia
construction period: 6 months
investment capital debt fraction: .40
annual interest on debt fraction: 9-1/2%
period of amortization of capital: 20 years
supervisory personnel: 2
operating personnel: 1
cost of fuel: 26¢/mcf
tax rate: 48%

We wish to compute the necessary revenues in ¢/mcf to realize a 15% rate of return on the investment.

Execution:

1. Compute Investment - M$

A. Pipeline

(reference LP-F-001 - Diameter for Minimum Investment computation). Present average cost for onshore pipelines
= $10,000/inch diameter/mile
= (7 miles)(18 inch)($10,000/inch diameter/mile)
= $1,260,000

B. Compressor Station

(1) (reference LP-F-001 - average present cost for a station is approximately $450/BHP)
(2) Compute approximate BHP requirements. (reference PL-E-001 - Cash Throw Computation)

To Execute: Load card PL-E-001, side 1 & 2

KEYSTROKES	OUTPUT
100 "A"	100.00
0.00 "A"	0.00
0.00 "A"	0.00
0.00 "A"	0.00

KEYSTROKES	OUTPUT
0.00 "A"	0.00
0.00 "A"	0.00
0.00 "A"	0.00
0.00 "A"	0.00
0.00 "A"	0.00
1.00 "A"	1.00
185.3 "A"	185.30 (psig)
985.3 "A"	985.30 (psig)

To Compute:

Press "E"	"5.0" (compression ratio)
	"107.27" BHP/MMcfd
	"10,727" (total BHP)
	"2,574.5" mcfd (fuel requirements)

10,727 BHP @ $450/HP = $4,827,150.00

C. Line Fill

$(I.D._2)^2(.372)(P_{avg})(.00528)$
$=(18)^2(.372)(363)(.00582)$
=231 mscf

For an exact computation refer to:

$$\frac{P_1V_1}{T_1z_1} = \frac{P_2V_2}{T_2z_2},$$

where z_2 is computed from GP-F-001 (Compressibility of Gases Computation)

231 mscf @ 26¢/mcf = $60,060.00

Summary Investment:

(a)	pipelines	$1,260,000.00
(b)	compressor station	4,827,150.00
(c)	line fill	60,060.00
	Total:	$6,147,210.00

2. Compute Operating Expenses per Year

A. Fixed Operating Expenses/Yr.

(1) supervisory, technical & clerical staff @$25M/man-yr. $40,000/yr.

(2)	operating personnel @$20M/man-yr.	$20,000/yr.
(3)	taxes and insurance @ 1.5%	20,000/yr.
(4)	pipeline & station maintenance @ 1.5%	92,000/yr.
(5)	miscellaneous @ .5% (leakage etc.)	31,000/yr.
	Total fixed expenses/yr.	$ 183,000.00

B. Variable Operating Expenses/Yr.

fuel and utilities - 2,574.5 mcfd (fuel requirements)

3. Royalties

Not applicable to this problem (reference introduction to PL-E-001)

4. BTU Penalty

Not applicable to this problem (reference introduction to PL-E-001)

5. Revenues

(Reference introduction to PL-E-001)

Select these revenues for economic evaluation:

5¢/mcf and 12¢/mcf

6. Compute Cash Throw

Load card PL-E-001, side 1 & 2

Store parameters in this order:

KEYSTROKES	OUTPUT
100 "A"	100.00
5 "A"	5.00
6,147,210 "A"	6,147,210.00
183,000 "A"	183,000.00
0.00 "A"	0.00
0.00 "A"	0.00
0.00 "A"	0.00
26 "A"	26.00
0.00 "A"	0.00
1.00 "A"	1.00
185.3 "A"	185.30
985.3 "A"	985.30

YR.	M$ REVENUES	M$ FIXED EXPENSES	M$ VARIABLE EXPENSES	CASH THROW	INTEREST	DEPRECI-ATION	CASH FLOW BEFORE TAX
15	1,825	183	244	1,398	70	252	1,076
16	1,825	183	244	1,398	58	252	1,088
17	1,825	183	244	1,398	47	252	1,099
18	1,825	183	244	1,398	35	252	1,111
19	1,825	183	244	1,398	23	252	1,123
20	1,825	183	244	1,398	12	252	1,134
TOTALS	36,500	3,660	4,880	27,960	2,451	5,040	20,469

Assume 48% taxes for 5¢/mcf revenues:

YEAR	CASH FLOW AFTER TAXES (M$)
1	438.24
2	443.52
3	449
4	454
5	460
6	466
7	471
8	477
9	482
10	488
11	493
12	500
13	506
14	510
15	516
16	522
17	527
18	533
19	539
20	544

8. Compute Internal Rate of Return

Load Card PL-E-005, side 1 & 2

Store parameters in this order:

KEYSTROKES	OUTPUT
6,147 "A"	6,147.00
438.24 "C"	438.24
443.52 "C"	443.52
449 "C"	449.00
454 "C"	454.00
460 "C"	460.00

KEYSTROKES	OUTPUT
466 "C"	466.00
471 "C"	471.00
477 "C"	477.00
482 "C"	482.00
488 "C"	488.00
493 "C"	493.00
500 "C"	500.00
506 "C"	506.00
510 "C"	510.00
516 "C"	516.00
522 "C"	522.00
527 "C"	527.00
533 "C"	533.00
539 "C"	539.00
544 "C"	544.00

To Compute:

Press "D" ____________ 4.74% (internal rate of return)

9. Compute Payout Period

Load card PL-E-006, side 1

Input/Compute:

KEYSTROKES	OUTPUT
6147 "A"	6,147.00 (M$)
4.74 "B"	4.74 (%)
438.24 "D"	-5,728.59
443.52 "D"	-5,324.31
449 "D"	-4,933.55
454 "D"	-4,556.32
460 "D"	-4,191.40
466 "D"	-3,838.45
471 "D"	-3,497.86
477 "D"	-3,168.54
482 "D"	-2,850.83
488 "D"	-2,543.72
493 "D"	-2,247.50
500 "D"	-1,960.68
506 "D"	-1,683.55
510 "D"	-1,416.86
516 "D"	-1,159.25
522 "D"	- 910.44
527 "D"	- 670.62
533 "D"	- 439.04
539 "D"	- 215.45
544 "D"	.002*

* Approximately 0 at year 20.

Rate of return on the investment: 4.74%

Payout in years: 20 years

Conclusion:

A 5¢/mcf revenue is unacceptable as a profitable venture.

YR.	M$ REVENUES	M$ FIXED EXPENSES	M$ VARIABLE EXPENSES	CASH THROW	INTEREST	DEPRECIATION	CASH FLOW BEFORE TAX
1	4,380	183	244	3,952	233	252	3,467
2	4,380	183	244	3,952	222	252	3,478
3	4,380	183	244	3,952	210	252	3,490
4	4,380	183	244	3,952	199	252	3,501
5	4,380	183	244	3,952	187	252	3,513
6	4,380	183	244	3,952	175	252	3,525
7	4,380	183	244	3,952	163	252	3,573
8	4,380	183	244	3,952	152	252	3,548
9	4,380	183	244	3,952	140	252	3,560
10	4,380	183	244	3,952	128	252	3,572
11	4,380	183	244	3,952	117	252	3,583
12	4,380	183	244	3,952	105	252	3,595
13	4,380	183	244	3,952	93	252	3,607
14	4,380	183	244	3,952	82	252	3,618
15	4,380	183	244	3,952	70	252	3,630
16	4,380	183	244	3,952	58	252	3,642
17	4,380	183	244	3,952	47	252	3,653
18	4,380	183	244	3,952	35	252	3,665
19	4,380	183	244	3,952	23	252	3,677
20	4,380	183	244	3,952	12	252	3,688
TOTALS	87,600	3,660	4,880	79,040	2,451	5,040	71,585

Assume 48% taxes for 12¢/mcf revenues:

YEAR	CASH FLOW AFTER TAXES (M$)
1	1802
2	1808
3	1814
4	1820
5	1826
6	1833
7	1839
8	1845
9	1851
10	1857
11	1863

YEAR	CASH FLOW AFTER TAXES (M$)
12	1849
13	1877
14	1881
15	1888
16	1894
17	1900
18	1905
19	1912
20	1918

Compute: Rate of Return on investment: 29.47%

Payout: 3.5 years (approximately)

Note: A graphical solution is now in order. Graph the rate of return (%) versus the revenue (¢/mcf).

4.74% = 5¢/mcf

29.47% = 12¢/mcf

Approximate revenues to yield 15% rate of return = 10¢/mcf, and a payout of approximately 4 years.

7.0 Normal Distribution Probability Computation for PERT and Tolerances Estimation

7.1 Introduction

The objectives of this program are to evaluate probabilistically, the outcome of random variables with respect to PERT networks and hydraulic equation tolerances.

7.2 Theoretical Formulae

(1) PERT:

$$\pm x = \frac{D - T}{S}$$

Where T = the sum of all expected activity duration of all activities along the critical path. ie. $te_1 + te_2 + te_3 + \ldots\ldots = T$

Where $te = \dfrac{to + 4tm + tp}{6}$

to = optimistic value
tm = most probable estimate
tp = pessimistic estimate

D = project due date to be selected at random for probability of occurance.

$$S = \frac{tp - to}{6}$$

$\pm x$ = the number of standard deviations by which D exceeds T.

Normal Distribution:

$$P = Q(x) = f(x)\ (b_1 t + b_2 t^2 + b_3 t^3 + b_4 t^4 + b_5 t^5) + e(x)$$

Where $f(x) = \dfrac{1}{\sqrt{2\pi}} e^{-(x^2/2)}$ $\quad |e(x)| < 7.5 \times 10^{-8}$

b_1 = .31938153
b_2 = -.356563782
b_3 = 1.781477937
b_4 = -1.821255978
b_5 = 1.330274429

$$Q(-x) = 1 - Q(x)$$

(2) Tolerances:

$$x' = \frac{D' - T'}{S'}$$

Where T' = expected "nominal" value of theoretical computation.

D' = random value of theoretical computation.

$S' = \dfrac{tp - to}{6}$

$\pm x'$ = the number of standard deviations by which D' exceeds T'.

7.3 Application of Theoretical Formulae

LP-F-007 - Normal Distribution Computation

Input:

±x value or ± x' value - n

Compute:

Percent probability of the random variable occurring - n%

STEP	INSTRUCTIONS	KEYS	OUTPUT UNITS
1.	load card LP-F-007, side 1 & 2		
2.	input optimistic estimate	fa	n
3.	input most likely estimate	fa	n
4.	input most pessimistic estimate	fa	n
5.	compute (STO 8) standard deviation	fb	n
6.	compute (STO 9) expected value	fc	n
7.	input random value to compute its probability of occurring	fd	± n
8.	compute probability	A	%n
9.	repeat steps 7 & 8 for other random variables.		

Example #1: LP-F-007 - Normal Distribution Computation

Given Conditions:

Using the Darcy-Weisbach flow equation a range of values is assigned as follows:

Nominal calculated theoretical value: 1,200,000 BBLS/day
Optimistic emperical value: 1,300,000 BBLS/day
Pessimistic emperical value: 1,100,000 BBLS/day

Note that a "nominal" flowrate computation will have intrinsic tolerances (ie. the equation itself, f factor, etc.) and extrinsic tolerances (ie., changes in s.g. and viscosity measurements).

Compute:
(1) the standard deviation about the nominal flowrate
(2) the expected flowrate
(3) a table of flowrate versus chances that a random flowrate will be met or bettered.

Execution: Load card LP-F-007, side 1 & 2.

	KEYSTROKES	OUTPUT
	1,300,000 "fa"	1,300,000.00
	1,200,000 "fa"	1,200,000.00
	1,100,000 "fa"	1,100,000.00
(standard deviation)	"fb"	33,333.33 - S'
(expected value)	"fc"	1,200,000.00 - T'
	1,000,000 "fd"	-6.00 - x'
	"A"	100.00 - %
	1,185,000 "fd"	-.45 - x'

KEYSTROKES	OUTPUT
"A"	67.36%
1,200,000 "fd"	0.00 - x'
"A"	50.00%

FLOWRATE BBLS/Day	Chances that Flowrate will be Met or Bettered
1,000,000	100%
1,185,000	67.36%
1,200,000	50%
1,215,000	33%
1,275,000	1%

Example #2: LP-F-007 - Normal Distribution Probability Computation

Given Conditions:

A manager has analyzed a project using PERT and computed the following along the critical path:
T = 20 months (expected date for completion)
S = 3 months (standard deviation along critical path)

Compute the probability that the project will be completed in the following number of months:

No. of months to Complete Project	Chances That Due date will be Met or Bettered
25	unknown
23	"
20	"
18	"
10	"

This computation is designed for a PERT network having a critical path much longer than any other path and/or many interconnections with a nearly parallel critical path

Execution: Load card LP-F-007, side 1 & 2.

KEYSTROKES	OUTPUT
3 "STO 8"	3.00 - S
20 "STO 9"	20.00 - T
25 "fd"	1.67 - x
"A"	4.78% - P
23 "fd"	1.00
"A"	15.87%
20 "fd"	0.00
"A"	50.00%
18 "fd"	-.67
"A"	75.00%
10 "Fd"	-3.33
"A"	100.00%

Random Due Date (months)	Chances that due date will be Met or Bettered
25	5%
23	16%
20	50%
18	75%
10	100%

References

7-1 Abramowitz et al, "Handbook of Mathematical Functions", National Bureau of Standards, 1968.

7-2 Wiest et al, "A Management Guide to PERT/CPM", Prentice-Hall, Inc. Englewood Cliffs, New Jersey. 1969

7-3 Miller, R.W., "How to Plan and Control with PERT", Harvard Business Review, March-April, 1962.

STEP	KEY ENTRY	KEY CODE
001	g LBLf a	32 25 11
	STO (i)	33 24
	f GSB 1	31 22 01
	h RTN	35 22
	f LBL 1	31 25 01
	f ISZ	31 34
	R/S	84
	h ST I	35 33
	GTO 1	22 01
010	h RTN	35 22
	g LBLf b	32 25 12
	RCL 0	34 00
	STO 5	33 05
	RCL 1	34 01
	STO 6	33 06
	RCL 2	34 02
	STO 7	33 07
	0	00
	STO 0	33 00
020	STO 1	33 01
	STO 2	33 02
	h ST I	35 33
	RCL 5	34 05
	RCL 7	34 07
	CHS	42
	+	61
	6	06
	÷	81
	STO 8	33 08
030	h RTN	35 22
	g LBLf c	32 25 13
	RCL 6	34 06
	4	04
	x	71
	RCL 5	34 05
	+	61
	RCL 7	34 07
	+	61
	6	06
040	÷	81
	STO 9	33 09
	h RTN	35 22
	g LBLf d	32 25 14
	RCL 9	34 09
	CHS	42
	+	61
	RCL 8	34 08
	÷	81
	h RTN	35 22
050	f LBL A	31 25 11
	STO A	33 11
	f $x < 0$	31 71
	GTO E	22 15
	g x^2	32 54
	2	02
	÷	81

STEP	KEY ENTRY	KEY CODE
	CHS	42
	g e^x	32 52
	STO B	33 12
060	h π	35 73
	2	02
	x	71
	f $\sqrt{x}$	31 54
	h 1/x	35 62
	RCL B	34 12
	x	71
	STO B	33 12
	f GSB B	31 22 12
	f GSB C	31 22 13
070	f GSB D	31 22 14
	STO E	33 15
	1	01
	0	00
	0	00
	x	71
	h RTN	35 22
	f LBL B	31 25 12
	RCL A	34 11
	.	83
080	2	02
	6	06
	1	01
	6	06
	4	04
	1	01
	9	09
	x	71
	1	01
	+	61
090	h 1/x	35 62
	STO C	33 13
	h RTN	35 22
	f LBL C	31 25 13
	.	83
	3	03
	1	01
	9	09
	3	03
	8	08
100	1	01
	5	05
	3	03
	RCL C	34 13
	x	71
	STO 0	33 00
	RCL C	34 13
	g x^2	32 54
	.	83
	3	03
110	5	05
	6	06
	5	05

LP-F-007

STEP	KEY ENTRY	KEY CODE
	6	06
	3	03
	7	07
	8	08
	2	02
	CHS	42
	x	71
120	STO 1	33 01
	RCL C	34 13
	3	03
	h y^x	35 63
	1	01
	.	83
	7	07
	8	08
	1	01
	4	04
130	7	07
	7	07
	9	09
	3	03
	7	07
	x	71
	STO 2	33 02
	RCL C	34 13
	4	04
	h y^x	35 63
140	1	01
	.	83
	8	08
	2	02
	1	01
	2	02
	5	05
	5	05
	9	09
	7	07
150	8	08
	CHS	42
	x	71
	STO 3	33 03
	RCL C	34 13
	5	05
	h y^x	35 63
	1	01
	.	83
	3	03
160	3	03
	0	00
	2	02
	7	07
	4	04
	2	02
	9	09
	x	71
	STO 4	33 04
	h RTN	35 22
170	f LBL D	31 25 14
	RCL 0	34 00
	RCL 1	34 01
	+	61
	RCL 2	34 02
	+	61
	RCL 3	34 03
	+	61
	RCL 4	34 04
	+	61
180	STO D	33 14
	RCL B	34 12
	x	71
	h RTN	35 22
	f LBL E	31 25 15
	RCL A	34 11
	CHS	42
	f GSB A	31 22 11
	CHS	42
	1	01
190	0	00
	0	00
	+	61
	h RTN	35 22

SECTION TWO

DESIGN

1.0 Liquids Pipeline Design

1.1 Introduction

The liquids design series contains seven programs for detailed design of liquid systems.

1.2 Theoretical Formulae

Formulae for this series is based primarily on the following formulae:

$$p = \frac{.323\ flsv^2}{D}$$

$$v = \frac{.0119Q}{I.D.^2}$$

$$R^* = \frac{.0119Q}{I.D.} = Re/7742$$

$$f_v = .00207\ (1/R^*)$$

$$f_i = .00814\ (R^*)^{-.3192}$$

$$f_{pt} = .0018 + .00662\ (1/R^*)^{.355} \simeq 4 \log \frac{Nre}{1.4126\ \sqrt{1/f_{pt}}}$$

$$\sqrt{1/f_{ft}} = 4\log(I.D./k_e) + 2.273$$

$$N_{Re}\ \text{critical} = 1.413 \left[\frac{3.7\ D}{k_e}\right]^{1/F_f} \sqrt{1/f_{ft}}$$

Limits of Application

$R^* < .135$ = Viscous flow

$R^* > .135 \lesssim .4$ = intermediate flow

$R^* > R^*$critical = fully trubulent flow

$R^* > .4 \lesssim R^*$critical = partially turbulent flow

Roughness Factor is assumed to be 750 micro inches

Drag Factor is assumed to be .95 (F_f)

$$\text{Brake Horsepower} = \frac{(PSI)(B/H)}{2450 \times \text{efficiency}}$$

Design Properties of Pipeline

$A_o = \frac{D\pi}{12}$ outside pipe surface area, ft.^{2}/L.F.

$A_i = \frac{d\pi}{12}$ inside pipe surface, ft.^{2}/L.F.

$A = \frac{(D^2 - d^2)\pi}{4}$ cross sectional metal area, in.^{2}

$A_p = \frac{d^2\pi}{4}$ cross sectional flow area, in.^{2}

$W_1 = 3.4A$ weight of pipe, lb./L.F.

$W_2 = 2.64W_1$ weight of pipe, Tons/mi.

$W_w = .433A_f$ weight of water filling, lb./L.F.

$r_g = \frac{\sqrt{D^2 + d^2}}{4}$ radius of gyration, in.

$= \sqrt{I/A_m}$

$I = A_m r_g^2 = .0491(D^4 - d^4)$ Moment of inertia, in.^{4}

$z = \frac{2I}{D} = .0982 \frac{(D^4 - d^4)}{D}$ section modulus, in.^{3}

Allowable Working Pressure of Pipeline

$$P = \frac{2ST}{D} \times F \times E \times T$$

Typical Product Specific Gravities and Absolute Viscosities @ 60° F.

	S.G.	z (centipoise)
Propane	.508	.120
Butane (average)	.573	.182
Gasoline 26 lb.	.655	.29
Gasoline 13 lb.	.679	.34
Gasoline 9 lb.	.724	.80
Fuel Oil No. 2 (average)	.850	4.06
Condensate	.600	.20
Kerosene	.820	2.00
Water	1.00	1.10
Crude Oil (approximate)	.840	15.00

Hydraulic Horsepower

Hydraulic Horsepower = BBLS. per day x PSIG x .000017

Hydraulic Horsepower = BBLS. per hr. x PSIG x .000408

Hydraulic Horsepower = GALS. per min. x PSIG x .000583

Velocity

feet per second = BBLS. per hour x .2859 ÷ $(I.D.")^2$

Rate of Flow

Gallons per minute = Barrels per day x .0292

Gallons per minute = Barrels per hour x .7

Density

Pressure, pounds per sq. in. per ft. height = 0.4326 x s.g.

Density, pounds per cubic foot = 62.2994 x s.g.

Density, pounds per gallon = 8.32823 x s.g.

Viscosity

Absolute Viscosity = 0.22SU - (180S ÷ U), centipoise

Kinematic Viscosity = 0.22U - (180 ÷ U), centistokes

Specific Gravity

Specific Gravity $60^{o}/60^{o}$ F = 141.5 ÷ (API^{o} + 131.5)

Pipe Linefill

$I.D.^2$ x .00097 = barrels per foot

$I.D.^2$ x 5.12913 = barrels per mile

$I.D.^2$ x 28.79793 = cubic feet per mile

$I.D.^2$ x 215.42439 = gallons per mile

$I.D.^2$ x 3.18709 = barrels per kilometer

$I.D.^2$ x 133.85769 = gallons per kilometer

1.3 Application of Theoretical Formulae

LP-D-001 - Basic Gathering/Transmission Pipeline Design

Input:

(1) flowrate - BBLS/Day
(2) I.D. - inches
(3) viscosity - C.S.
(4) specific gravity - n
(5) length of line - miles
(6) No. of pump stations - n
(7) final elevation - feet
(8) initial elevation - feet
(9) suction pressure at terminal - psi
(10) specified minimum yield of pipe - psi
(11) wall thickness of pipeline - inches
(12) O.D. of pipeline - inches
(13) construction type design factor - n

Computes:

(1) Reynolds Number modified (Re ÷ 7742)
(2) friction factor
(3) pressure drop per mile
(4) total line pressure required
(5) total discharge pressure/station
(6) net pump discharge pressure/station
(7) BHP/station
(8) maximum allowable working pressure
(9) total BHP for all stations
(10) pressure drop gradient/mile

STEP	INSTRUCTIONS	KEYS	OUTPUT UNITS
1.	load side 1 & 2		
2.	input flowrate	A	BBLS/Day
3.	input I.D.	A	inches
4.	input viscosity	A	centistokes
5.	input specific gravity	A	n
6.	input length of line	A	miles
7.	input no. of pump stations	A	n
8.	input final elevation	A	feet
9.	input initial elevation	A	feet
10.	input suction pressure at terminal	A	psi
11.	input specified minimum yield of pipe	A	psi
12.	input wall thickness of pipeline	A	inches
13.	input O.D. of pipeline	A	inches
14.	input construction type design factor	A	n

STEP	INSTRUCTIONS	KEYS	OUTPUT UNITS
15.	Compute: (1) Reynolds number modified (2) friction factor (3) pressure drop per mile	B	
16.	Compute: (1) total line pressure (2) total discharge pressure (3) net pump discharge pressure (4) BHP/station (efficiency = .7) (5) MAWP of pipeline (6) total BHP for n stations	C	
17.	Compute: (1) pressure gradient (2) location of pump stations	D	
18.	Options: (1) clear registers for new case (2) pressure drop per mile for Q in B/H; input B/H	 fa fb	 ΔPSI/mi.

Example LP-D-001

An LPG transmission system is to be designed for a capacity of 52,910 BPD, the parameters are as follows:

flowrate: 52,910 BPD
I.D. of pipeline: 8.125"
viscosity: .31 C.S.
specific gravity: .582
length: 82.3 miles
no. of pump stations: 3
final elevation: 625'
initial elevation: 600'
suction (terminal) pressure: 450 psi
specified minimum yield of pipe: 35,000 psi
W.T.: .250"
O.D.: 8.625"
construction type design factor: .72

Compute pressure drop and BHP parameters for this system.
Load card LP-D-001, side 1 & 2,
Store parameters in this order:

KEYSTROKES	OUTPUTS
52,910 "A"	52,910.0000
8.125 "A"	8.1250
.31 "A"	.3100

KEYSTROKES	OUTPUTS
.582 "A"	.5820
82.3 "A"	82.3000
3 "A"	3.0000
625 "A"	625.0000
600 "A"	600.0000
450 "A"	450.0000
35,000 "A"	35,000.0000
.250 "A"	.2500
8.625 "A"	8.6250
.72 "A"	.7200

Compute: Reynolds no. modified; friction factor; ΔPSI/mi.

Execution:

Press B	"249.9768"
	" .0027"
	" 30.3001"

Option to obtain log-log graph (B/H versus ΔPSI/mi.)

500 B/H "fb"	"56.6948"
	" .0034"
	" 1.9273"
1000 B/H "fb"	"113.3896"
	" .0030"
	" 6.9236"
1500 B/H "fb"	"170.0844"
	" .0029"
	" 14.7284"

Compute: Total line pressure; total discharge pressure; net pump discharge pressure; BHP/station; MAWP of pipeline; total BHP for all stations.

52,910÷24 "fb"	"249.9768"
	" .0027"
	" 30.3001"
"C"	"2493.6946"
	"1287.5317"
	" 837.5317"
	"1076.6230"
	"1460.8696"
	"1076.6230" (total BHP with one station)
"C"	"2493.6946"
	"1287.5317"
	" 837.5317"
	"1076.6230"
	"1460.8696"
	"2153.2460" (total BHP with two stations)

KEYSTROKES	OUTPUTS
"C"	"2493.6946"
	"1287.5317"
	" 837.5317"
	"1076.6230"
	"1460.8696"
	"3229.8690" (total BHP with 3 stations)

Compute: Pressure Drop Gradient

Input elevation at M.P. 81.3: (82.3 - 1)

625 "D"	"1.0" M.P.
	"81.3" M.P. actual
	"480.3" PSI
	"1287.5317" max. discharge pressure of pump station - a constant
625 "D"	"2.0"
	"80.3"
	"510.6001"
	"1287.5317"
625 "D"	"3.0"
	"79.3"
	"540.9002"
	"1287.5317"
625 "D"	"4.0"
	"78.3"
	"571.2002"
	"1287.5317"
625 "D"	"5.0"
	"77.3"
	"601.5003"
	"1287.5317"
625 "D"	"6.0"
	"76.3"
	"631.8003"
	"1287.5317"
625 "D"	"7.0"
	"75.3"
	"662.1004"
	"1287.5317"
625 "D"	"8.0"
	"74.3"
	"692.4005"
	"1287.5317"

KEYSTROKES	OUTPUTS
625 "D"	"9.0"
	"73.3"
	"722.7005"
	"1287.5317"
625 "D"	"10.0"
	"72.3"
	"753.0006"
	"1287.5317"
625 "D"	"11.0"
	"71.3"
	"783.3006"
	"1287.5317"
625 "D"	"12.0"
	"70.3"
	"813.6007"
	"1287.5317"
625 "D"	"13.0"
	"69.3"
	"843.9007"
	"1287.5317"
625 "D"	"14.0"
	"68.3"
	"874.2008"
	"1287.5317"
625 "D"	"15.0"
	"67.3"
	"904.5008"
	"1287.5317"
625 "D"	"16.0"
	"66.3"
	"934.8009"
	"1287.5317"
625 "D"	"17.0"
	"65.3"
	"965.1010"
	"1287.5317"
625 "D"	"18.0"
	"64.3"
	"995.4010"
	"1287.5317"

KEYSTROKES	OUTPUTS
625 "D"	"19.0"
	"63.3"
	"1025.7011"
	"1287.5317"
625 "D"	"20.0"
	"62.3"
	"1056.5317"
	"1287.5317"
625 "D"	"21.0"
	"61.3"
	"1086.3012"
	"1287.5317"
625 "D"	"22.0"
	"60.3"
	"1116.6012"
	"1287.5317"
625 "D"	"23.0"
	"59.3"
	"1146.9013"
	"1287.5317"
625 "D"	"24.0"
	"58.3"
	"1177.2014"
	"1287.5317"
625 "D"	"25.0"
	"57.3"
	"1207.5014"
	"1287.5317"
625 "D"	"26.0"
	"56.3"
	"1237.8015"
	"1287.5317"
625 "D"	"27.0"
	"55.3"
	"1268.1015"
	"1287.5317"
625 "D"	"28.0"
	"54.3"
	"1298.4016"
	"5.7386" ΔPSI/1000'

(pump discharge pressure is exceeded - station required)

```
  1298.4016
-    5.7386
  1292.6630   ΔPSI/1000'
-    5.7386
  1286.9244   ΔPSI/1000'
```

Location of Pump Station No. 3

54.3 + .4 = 54.7 Mile Post
Suction pressure = 450 PSI
Discharge pressure = 1288 PSI

Location of Pump Station No. 2

27.64 Mile Post
Suction pressure = 450 PSI
Discharge pressure = 1288 PSI

Location of Pump Station No. 1

0.00 Mile Post
Suction pressure = 450 PSI
Discharge pressure = 1288 PSI

Pump Station at 54.7

	KEYSTROKES	OUTPUT
(suction pressure)	450 "STO E"	450 (start with elevation of new station)
	625 "D"	"29.0"
		"53.3"
		"480"
		"1287.5317"
	625 "D"	"30.0"
		"52.3"
		"510.6001"
		"1287.5317"

LP-D-001

STEP	KEY ENTRY	KEY CODE
001	f LBL A	31 25 11
	STO (i)	33 24
	f GSB 1	31 22 01
	h RTN	35 22
	f LBL 1	31 25 01
	f ISZ	31 34
	R/S	84
	h ST I	35 33
	GTO 1	22 01
010	h RTN	35 22
	f LBL B	31 25 12
	.	83
	0	00
	1	01
	1	01
	9	09
	RCL 0	34 00
	x	71
	RCL 1	34 01
020	÷	81
	RCL 2	34 02
	÷	81
	DSP 4	23 04
	f -x-	31 84
	STO A	33 11
	.	83
	4	04
	g x≤y	32 71
	GTO 2	22 02
030	RCL A	34 11
	.	83
	1	01
	3	03
	5	05
	g x≤y	32 71
	GTO 3	22 03
	h x≷y	35 52
	h 1/x	35 62
	.	83
040	0	00
	0	00
	2	02
	0	00
	7	07
	x	71
	f -x-	31 84
	STO B	33 12
	GTO 5	22 05
	f LBL 3	31 25 03
050	h x≷y	35 52
	.	83
	3	03
	1	01
	9	09
	2	02
	CHS	42
	h y^x	35 63
	.	83
	0	00
060	0	00
	8	08
	1	01
	4	04
	x	71
	STO B	33 12
	f -x-	31 84
	GTO 5	22 05
	f LBL 2	31 25 02
	h x≷y	35 52
070	h 1/x	35 62
	.	83
	3	03
	5	05
	5	05
	h y^x	35 63
	.	83
	0	00
	0	00
	6	06
080	6	06
	2	02
	x	71
	.	83
	0	00
	0	00
	1	01
	8	08
	+	61
	f -x-	31 84
090	STO B	33 12
	f LBL 5	31 25 05
	RCL B	34 12
	RCL 3	34 03
	x	71
	.	83
	2	02
	4	04
	1	01
	x	71
100	RCL 0	34 00
	g x^2	32 54
	x	71
	RCL 1	34 01
	5	05
	h y^x	35 63
	÷	81
	f -x-	31 84
	STO C	33 13
	h RTN	35 22
110	g LBLf b	32 25 12
	2	02
	4	04

STEP	KEY ENTRY	KEY CODE
	x	71
	STO 0	33 00
	GTO B	22 12
	h RTN	35 22
	g LBLf a	32 25 11
	0	00
	h ST I	35 33
120	f CL REG	31 43
	h RTN	35 22
	f LBL C	31 25 13
	RCL 4	34 04
	RCL C	34 13
	x	71
	f -x-	31 84
	f GSB 4	31 22 04
	RCL 5	34 05
	x	71
130	+	61
	RCL 8	34 08
	STO E	33 15
	RCL 5	34 05
	x	71
	+	61
	RCL 5	34 05
	÷	81
	STO D	33 14
	f -x-	31 84
140	RCL 8	34 08
	-	51
	f -x-	31 84
	RCL 0	34 00
	2	02
	4	04
	÷	81
	x	71
	1	01
	7	07
150	1	01
	5	05
	÷	81
	f -x-	31 84
	f P≷S	31 42
	STO +5	33 61 05
	f P≷S	31 42
	RCL 9	34 09
	f P≷S	31 42
	RCL 0	34 00
160	x	71
	2	02
	x	71
	RCL 2	34 02
	x	71
	RCL 1	34 01
	÷	81
	f -x-	31 84
	RCL 5	34 05

STEP	KEY ENTRY	KEY CODE
	f -x-	31 84
170	f P≷S	31 42
	h RTN	35 22
	f LBL 4	31 25 04
	RCL 6	34 06
	RCL 7	34 07
	-	51
	.	83
	4	04
	3	03
	3	03
180	x	71
	RCL 3	34 03
	x	71
	h RTN	35 22
	f LBL D	31 25 14
	STO 7	33 07
	f GSB 7	31 22 07
	f GSB 4	31 22 04
	RCL C	34 13
	+	61
190	RCL E	34 15
	+	61
	STO E	33 15
	RCL 7	34 07
	STO 6	33 06
	RCL E	34 15
	f -x-	31 84
	RCL D	34 14
	g x>y	32 81
	h RTN	35 22
200	GTO 6	22 06
	f LBL 7	31 25 07
	f P≷S	31 42
	1	01
	STO +6	33 61 06
	RCL 6	34 06
	f -x-	31 84
	f P≷S	31 42
	RCL 4	34 04
	h x≷y	35 52
210	-	51
	f -x-	31 84
	h RTN	35 22
	f LBL 6	31 25 06
	RCL C	34 13
	5	05
	.	83
	2	02
	8	08
	÷	81
220	f -x-	31 84
	h RTN	35 22

LP-D-002 - Basic Design Parameters

Input:

(1) volume (option: input/output) - BBLS/Day
(2) internal diameter (option: input/output) - inches
(3) maximum design operating pressure - PSI
(4) outside diameter - inches
(5) specified minimum yield - PSI
(6) construction design factor - n

Computes:

(1) allowable wall thickness based on Barlow's formula - inches
(2) calculation of regularly commercially available wall thickness - inches
(3) Design parameters of pipe
    (a) inside diameter - inches
    (b) outside surface area - $ft.^2$/L.F.
    (c) inside surface area - $ft.^2$/L.F.
    (d) cross sectional metal area - $in.^2$
    (e) cross sectional flow area - $in.^2$
    (f) weight of pipe - lb./L.F.
    (g) weight of pipe - Tons/mi.
    (h) weight of water fill - lb./L.F.
    (i) radius of gyration - inches
    (j) moment of inertia - $in.^4$
    (k) section modulus - $in.^3$

STEP	INSTRUCTIONS	KEYS	OUTPUT UNITS
1.	load card LP-D-002		
2.	input volume (option to compute)	B	BBLS/Day
3.	input internal diameter (option to compute)	A	inches
4.	input maximum design operating pressure	fc	PSI
5.	input outside pipe diameter	fc	inches
6.	input specified minimum yield of pipe	fe	PSI
7.	input construction type design factor	fc	n
8.	Compute:		
	(1) calculate required wall thickness	C	inches
	(2) prepare for data card	fd	19.0
	(3) press "g merge" and enter appropriate data card	g merge	
	(4) press "R/S" to reset search program	R/S	
	(5) search for commercially available wall thickness	D	inches

STEP	INSTRUCTIONS	KEYS	OUTPUT UNITS
	(a) Option to recompute commercially available wall thickness for the same outside diameter. Go to steps 4, 5, 6, 7, 8(1), and press fa.		
	Press D for commercially available wall thickness with these parameters.	D	
	(b) Option to recompute commercially available wall thickness for a different outside diameter. Go to steps 4, 5, 6, 7, 8(1), 8(2), 8(3), 8(4), 8(5).		
9.	Options:		
	(1) Compute design parameters as listed under Computes: (3), a-k.	E	
	(2) Option to compute design parameters for different I.D.		
	(a) input O.D.	ENTER ↑	inches
	(b) input wall thickness	fe	inches
	(c) compute design parameters as listed under Computes: (3) a-k.	E	

Example LP-D-002

Compute basic design parameters for an LPG system with a capacity of 52,910 B/D. The maximum design operating pressure is 1429 PSI, SMY is 35,000 PSI, and the construction design factor is .72.

Execution: Load card LP-D-002, side 1 & 2, Initial Approximation

KEYSTROKES	OUTPUTS
52,910 B/D "B"	52,910.000
Press "A"	10.516 (internal diameter @ 5.7 ft/sec)

Option to Input I.D.

KEYSTROKES	OUTPUTS
10.516 "A"	10.516
Press "B"	52,910.000 (flowrate)

Note: A 12" line appears to be satisfactory for initial consideration.

KEYSTROKES	OUTPUTS
1429 "fc"	1,429.000
12.750 "fc"	12.750
35,000 "fc"	35,000.000
.72 "fc"	.720

Compute: minimum wall thickness

KEYSTROKES	OUTPUTS
"C"	" .362"
	"12.750"
"fd"	"19.0"
Press "g merge" & insert data card G/L-WT-12/16	"19.0" (data card with 12" W.T.)
Press "R/S"	" 0.00"
Press "D"	" .375"

Option: To search for new commercially available wall thicknesses for the same outside diameter, but with different parameters.

KEYSTROKES	OUTPUTS
1200 PSI "fc"	1,200.000 : Maximum design pressure
12.750" "fc"	12.750 : O.D.
42,000 "fc"	42,000.000 : SMY
.72 "fc"	.720 : F
Press "C"	".253"
	12.750
Press "fa"	0.000
Press "D"	.281

Option: To recompute commercially available wall thicknesses for a different outside diameter.

KEYSTROKES	OUTPUTS
1429 "fc"	1,429.000
6.625 "fc"	6.625
35,000 "fc"	35,000.000
.72 "fc"	.720
Press "C"	" .188"
	6.625
Press "fd"	19.000
Press "g merge" and load card G/L-WT-4/6	19.000
Press "R/S"	0.000
Press "D"	.188

Option: To compute design properties of pipeline

Press "E"	" 6.249"	I.D. inches
	" 1.734"	A_o ft.^{2}/L.F.
	" 1.636"	A_i ft.^{2}/L.F.
	" 3.802"	A_m in.^{2}
	"30.670"	A_f in.^{2}
	"12.926"	W_1 lb./L.F.
	"34.125"	W_2 TONS/mi.
	"13.280"	W_3 lb./L.F.
	" 2.277"	r_g in.
	"19.708"	I in.^{4}
	" 5.950"	Z in.^{3}

Option: To compute design properties for any O.D. & W.T.

12.750 ENTER↑	12.750
.375 "fe"	12.750
Press "E"	" 12.000"
	" 3.338"
	" 3.142"
	" 14.579"
	"113.097"
	" 49.568"
	"130.861"
	" 48.971"
	" 4.377"
	"279.335"
	" 43.817"

Note: ERW pipe is available up to and including 16" O.D. Double submerged arc and spiral weld are available for 16" O.D. and beyond.

ERW pipe wall thickness is to be limited to .375 inches.

The approximation for flowrates and internal diameter are based on a fluid flow average velocity of 5.69 feet/second.

Data Cards: To be Used With Program LP-D-002

G/L-WT-4/6	-	4.5 & 6.625 inch
G/L-WT-8/10	-	8.625 & 10.750 inch
G/L-WT-12/16	-	12.750 & 16.0 inch
G/L-WT-20/24	-	20.0 & 24.0 inch
G/L-WT-26/30	-	26.0 & 30.0 inch

G/L-WT-4/6 - Storage Registers

0 - 4.50	S0 - 6.625
1 - .125	S1 - .156
2 - .156	S2 - .188
3 - .188	S3 - .219
4 - .219	S4 - .250
5 - .237	S5 - .280
6 - .250	S6 - .312
7 - .281	S7 - .375
8 - .312	S8 - .432
9 - .337	

G/L-WT-8/10 - Storage Registers

0 - 8.625	S0 - .375	A - .344
1 - .156	S1 - .438	B - .365
2 - .188	S2 - .500	C - .438
3 - .203	S3 - 10.750	D - .500
4 - .219	S4 - .188	
5 - .250	S5 - .203	
6 - .277	S6 - .219	
7 - .312	S7 - .250	
8 - .322	S8 - .279	
9 - .344	S9 - .307	

G/L-WT-12/16 - Storage Registers

0 - 12.750	S0 - .406	A - .438
1 - .188	S1 - .438	B - .500
2 - .203	S2 - .500	C - .625
3 - .219	S3 - 16.00	
4 - .250	S4 - .219	
5 - .281	S5 - .250	
6 - .312	S6 - .281	
7 - .330	S7 - .312	
8 - .344	S8 - .344	
9 - .375	S9 - .375	

G/L-WT-20/24 - Storage Registers

0 - 20.000	S0 - .750	A - .750
1 - .250	S1 - 24.000	
2 - .281	S2 - .281	
3 - .312	S3 - .312	
4 - .344	S4 - .344	
5 - .375	S5 - .375	
6 - .406	S6 - .406	
7 - .438	S7 - .438	
8 - .500	S8 - .500	
9 - .625	S9 - .625	

G/L-WT-26/30 - Storage Registers

0 - 26.00	S0 - 30.00
1 - .281	S1 - .312
2 - .312	S2 - .344
3 - .344	S3 - .375
4 - .375	S4 - .406
5 - .406	S5 - .438
6 - .438	S6 - .500
7 - .500	S7 - .625
8 - .625	S8 - .750
9 - .750	

STEP	KEY ENTRY	KEY CODE
001	f LBL A	31 25 11
	STO A	33 11
	h F? 3	35 71 03
	h RTN	35 22
	f GSB 2	31 22 02
	RCL B	34 12
	x	71
	f $\sqrt{x}$	31 54
	STO A	33 11
010	f GSB 5	31 22 05
	RCL A	34 11
	h RTN	35 22
	f LBL B	31 25 12
	STO B	33 12
	h F? 3	35 71 03
	h RTN	35 22
	f GSB 2	31 22 02
	RCL C	34 13
	h 1/x	35 62
020	RCL A	34 11
	g x^2	32 54
	x	71
	STO B	33 12
	f GSB 5	31 22 05
	RCL B	34 12
	h RTN	35 22
	f LBL C	31 25 13
	RCL A	34 11
	RCL B	34 12
030	x	71
	RCL C	34 13
	÷	81
	RCL D	34 14
	÷	81
	2	02
	÷	81
	STO D	33 14
	f -x-	31 84
	f GSB 5	31 22 05
040	RCL B	34 12
	STO E	33 15
	h RTN	35 22
	g LBLf c	32 25 13
	STO (i)	33 24
	f GSB 1	31 22 01
	h RTN	35 22
	f LBL 1	31 25 01
	f ISZ	31 34
	R/S	84
050	h ST I	35 33
	GTO 1	22 01
	h RTN	35 22
	f LBL D	31 25 14
	RCL E	34 15
	RCL (i)	34 24
	-	51

STEP	KEY ENTRY	KEY CODE
	f x=0	31 51
	GTO 4	22 04
	f GSB 3	31 22 03
060	GTO D	22 14
	f LBL 4	31 25 04
	f GSB 3	31 22 03
	RCL D	34 14
	RCL (i)	34 24
	g x≤y	32 71
	GTO 4	22 04
	STO B	33 12
	f GSB 5	31 22 05
	RCL B	34 12
070	h RTN	35 22
	f LBL 3	31 25 03
	h RC I	35 34
	1	01
	+	61
	h ST I	35 33
	h RTN	35 22
	f LBL 2	31 25 02
	.	83
	0	00
080	0	00
	2	02
	0	00
	9	09
	STO C	33 13
	RCL C	34 13
	h RTN	35 22
	g LBLf d	32 25 14
	1	01
	9	09
090	h ST I	35 33
	R/S	84
	f GSB 6	31 22 06
	h RTN	35 22
	f LBL 5	31 25 05
	2	02
	0	00
	h ST I	35 33
	h RTN	35 22
	g LBLf a	32 25 11
100	f GSB 6	31 22 06
	h RTN	35 22
	f LBL 6	31 25 06
	0	00
	h ST I	35 33
	h RTN	35 22
	f LBL E	31 25 15
	RCL E	34 15
	RCL B	34 12
	2	02
110	x	71
	-	51
	f -x-	31 84

STEP	KEY ENTRY	KEY CODE
	STO B	33 12
	RCL E	34 15
	h π	35 73
	x	71
	1	01
	2	02
	÷	81
120	f -x-	31 84
	RCL B	34 12
	h π	35 73
	x	71
	1	01
	2	02
	÷	81
	f -x-	31 84
	RCL E	34 15
	g x^2	32 54
130	RCL B	34 12
	g x^2	32 54
	-	51
	h π	35 73
	x	71
	4	04
	÷	81
	f -x-	31 84
	STO D	33 14
	RCL B	34 12
140	g x^2	32 54
	h π	35 73
	x	71
	4	04
	÷	81
	f -x-	31 84
	STO C	33 13
	RCL D	34 14
	3	03
	.	83
150	4	04
	x	71
	f -x-	31 84
	2	02
	.	83
	6	06
	4	04
	x	71
	f -x-	31 84
	RCL C	34 13
160	.	83
	4	04
	3	03
	3	03
	x	71
	f -x-	31 84
	RCL E	34 15
	g x^2	32 54
	RCL B	34 12

STEP	KEY ENTRY	KEY CODE
	g x^2	32 54
170	+	61
	f $\sqrt{x}$	31 54
	4	04
	÷	81
	f -x-	31 84
	g x^2	32 54
	RCL D	34 14
	x	71
	f -x-	31 84
	2	02
180	x	71
	RCL E	34 15
	÷	81
	f -x-	31 84
	0	00
	STO B	33 12
	STO D	33 14
	STO C	33 13
	h RTN	35 22
	g LBLf e	32 25 15
190	STO B	33 12
	h $x \gtrless y$	35 52
	STO E	33 15
	h RTN	35 22

LP-D-003 - Flowrate Computation

Input:

(1) 0.00 - constant
(2) ΔPSI/mi. - PSI/mi.
(3) 0.433 - constant
(4) I.D. - inches
(5) viscosity - centistokes
(6) specific gravity

Computes:

(1) approximate flow rate for a given velocity
(2) flowrate in barrels per day

STEP	INSTRUCTIONS	KEY	OUTPUT UNITS
1.	load card LP-D-005		
2.	load card LP-D-003		
3.	input 0.00	A	0.00
4.	input pressure drop per mile	A	PSI/mi.
5.	input 0.433	A	0.433
6.	input internal diameter	A	inches
7.	input viscosity	A	centistokes
8.	input specific gravity	A	
9.	Options:		
	(1) approximate flowrate for a given velocity.		
	Input velocity ft./sec.	D	BBLS/Day
	(2) elevation factor.		
	Input length (miles) of pipeline	STO 0	miles
	Input initial elevation (ft)	ENTER↑	feet
	Input final elevation (ft)	fb	ΔPSI/mi.
10.	Compute Flowrate:		
	(1) Input velocity ft./sec.	D	BBLS/Day
		E	BBLS/Day
	(2) new case; I register is set to zero.	C	0.00

Example LP-D-003

What is the capacity of a 10 inch gasoline pipeline 50 miles long, loss of head (pressure drop) 792 lbs. per square inch, gravity 60° API and viscosity .555 centpoise.

(1) $\Delta PSI/mi. = \dfrac{729 \text{ lbs./in.}^2}{50 \text{ mi.}} = 15.84$

(2) 60° API = 141.5 ÷ (131.5 + 60° API) = .739 s.g.

(3) centistokes = $\dfrac{.555 \text{ cp}}{.739}$ = .75 centistokes

Load card LP-D-003, side 1 & 2,
Store parameters in this order:

KEYSTROKES	OUTPUTS
0.00 "A"	0.000
15.84 "A"	15.840
.433 "A"	.433
10.00 "A"	10.000
.75 "A"	.750
.739 "A"	.739

Compute: Approximate flowrate at 7 ft./sec; actual velocity; actual flowrate.

Execution:

7.00"D"(option)	58,823.53
"E"	6.30 ft./sec.
	52,941.76 B/D

Repeat above example except initial elevation = 100 ft.
final elevation = 0 ft.

"C"	" 0.00"
50 mi. "STO 0"	" 50.00"
100 ENTER↑	" 100.00"
0 "fb"	" + .64" PSI/mi.
0.00 "A"	" 0.00"
15.84 + .64 =	
16.48 "A"	" 16.48"
.433 "A"	" .433"
10 "A"	" 10.00"
.75 "A"	" .75"
.739 "A"	" .739"
7 "D"	"58,823.53" B/D
"E"	" 6.44" ft./sec.
	"54,089.48" B/D

STEP	KEY ENTRY	KEY CODE
001	f LBL A	31 25 11
	STO (i)	33 24
	f GSB 3	31 22 03
	h RTN	35 22
	f LBL 3	31 25 03
	f ISZ	31 34
	R/S	84
	h ST I	35 33
	GTO 3	22 03
010	h RTN	35 22
	f LBL E	31 25 15
	f FIX	31 23
	GTO B	22 12
	g LBLf a	32 25 11
	RCL B	34 12
	GTO 0	22 00
	f LBL 6	31 25 06
	RCL 0	34 00
	f GSB 1	31 22 01
020	STO B	33 12
	f LBL 0	31 25 00
	RCL A	34 11
	RCL 0	34 00
	STO A	33 11
	-	51
	RCL D	34 14
	RCL B	34 12
	STO D	33 14
	-	51
030	÷	81
	x	71
	STO -0	33 51 00
	RCL 0	34 00
	h PAUSE	35 72
	÷	81
	f RND	31 24
	f x≠0	31 61
	GTO 6	22 06
	RCL 0	34 00
040	g GSB e	32 22 15
	f -x-	31 84
	h x≷y	35 52
	h RTN	35 22
	f LBL B	31 25 12
	.	83
	0	00
	1	01
	f %	31 82
	STO C	33 13
050	2	02
	÷	81
	-	51
	STO A	33 11
	STO 0	33 00
	f GSB 1	31 22 01
	STO D	33 14
	RCL A	34 11
	RCL C	34 13
	+	61
060	STO 0	33 00
	f GSB 1	31 22 01
	STO B	33 12
	RCL D	34 14
	-	51
	RCL C	34 13
	÷	81
	GTO f a	22 31 11
	f LBL 1	31 25 01
	RCL 9	34 09
070	.	83
	0	00
	0	00
	2	02
	0	00
	7	07
	-	51
	f x=0	31 51
	GTO 2	22 02
	7	07
080	ENTER ↑	41
	0	00
	÷	81
	h RTN	35 22
	f LBL 2	31 25 02
	RCL 1	34 01
	RCL 3	34 03
	5	05
	h y^x	35 63
	x	71
090	RCL 7	34 07
	÷	81
	RCL 5	34 05
	÷	81
	f $\sqrt{x}$	31 54
	f P≷S	31 42
	STO 5	33 05
	f GSB 9	31 22 09
	h 1/x	35 62
	f $\sqrt{x}$	31 54
100	STO 9	33 09
	RCL 5	34 05
	x	71
	f P≷S	31 42
	RCL 0	34 00
	-	51
	h RTN	35 22
	f LBL 9	31 25 09
	f P≷S	31 42
	RCL 6	34 06
110	RCL 0	34 00
	x	71
	RCL 3	34 03

STEP	KEY ENTRY	KEY CODE
	÷	81
	RCL 4	34 04
	÷	81
	f P≷S	31 42
	STO 9	33 09
	.	83
	4	04
120	g x≤y	32 71
	GTO 5	22 05
	RCL 9	34 09
	f P≷S	31 42
	RCL 8	34 08
	g x≤y	32 71
	GTO 7	22 07
	h x≷y	35 52
	h 1/x	35 62
	RCL 9	34 09
130	x	71
	f P≷S	31 42
	h RTN	35 22
	f LBL 5	31 25 05
	RCL 9	34 09
	GTO 8	22 08
	g LBLf d	32 25 14
	g x≤y	32 71
	GTO f c	22 31 13
	h x≷y	35 52
140	h 1/x	35 62
	f P≷S	31 42
	RCL 2	34 02
	h y^x	35 63
	RCL 3	34 03
	x	71
	RCL 4	34 04
	+	61
	h RTN	35 22
	f LBL 8	31 25 08
150	f P≷S	31 42
	RCL 3	34 03
	f P≷S	31 42
	RCL 8	34 08
	÷	81
	RCL 7	34 07
	x	71
	STO 9	33 09
	f LOG	31 53
	4	04
160	x	71
	RCL 9	34 09
	RCL 6	34 06
	h y^x	35 63
	x	71
	1	01
	.	83
	4	04
	x	71
	7	07
170	7	07
	4	04
	2	02
	÷	81
	f P≷S	31 42
	GTO f d	22 31 14
	h RTN	35 22
	f LBL 7	31 25 07
	h x≷y	35 52
	f P≷S	31 42
180	RCL 0	34 00
	h y^x	35 63
	RCL 1	34 01
	x	71
	h RTN	35 22
	g LBLf c	32 25 13
	f P≷S	31 42
	RCL 9	34 09
	f LOG	31 53
	4	04
190	x	71
	g x^2	32 54
	h 1/x	35 62
	f -x-	31 84
	h RTN	35 22
	g LBLf b	32 25 12
	-	51
	RCL 5	34 05
	RCL 2	34 02
	x	71
200	x	71
	RCL 0	34 00
	÷	81
	STO +1	33 61 01
	h RTN	35 22
	f LBL C	31 25 13
	0	00
	h ST I	35 33
	h RTN	35 22
	g LBLf e	32 25 15
210	RCL 6	34 06
	RCL 0	34 00
	x	71
	RCL 3	34 03
	g x^2	32 54
	÷	81
	h RTN	35 22
	f LBL D	31 25 14
	RCL 3	34 03
	g x^2	32 54
220	x	71
	RCL 6	34 06
	÷	81
	h RTN	35 22

LP-D-004 - Pressure Drop Computation

Input:

(1) specific gravity
(2) flowrate - BBLS/Day
(3) internal diameter - inches
(4) viscosity - c.s.
(5) length of pipeline - feet

Computes:

(1) velocity - ft./sec.
(2) Reynolds No. modified
(3) Reynolds No. critical
(4) type of flow
(5) ΔPSI/L
(6) BHP/L; -GPM-

STEP	INSTRUCTIONS	KEYS	OUTPUT UNITS
1.	load card LP-D-005		
2.	load card LP-D-004		
3.	input specific gravity	A	
4.	input flowrate	A	BBLS/Day
5.	input internal diameter	A	inches
6.	input viscosity	A	centistokes
7.	input length	A	feet
8.	Option to input flowrate in GPM	fb	BBLS/Day
	Option to clear for new case	fa	0.00
9.	Compute:	B	
	(1) velocity of flow		ft./sec.
	(2) Reynolds No. modified		
	(3) Reynolds No. critical		
	(4) type of flow, where;		
	1 = viscous		
	2 = intermediate		
	3 = partially turbulent		
	4 = fully trubulent		
	(5) pressure drop per Input: length		
10.	Options:		
	(1) clear all registers	fa	0.00
	(2) compute BHP/L	D	gpm BHP/L

Example LP-D-004

What is the pressure drop due to friction per mile in a ten inch gasoline line 50 miles long when discharge capacity is 2,205.9 BBLS/hour, the gravity of the gasoline is .739 and the viscosity is .75 centistokes.

Load cards LP-D-005 and LP-D-004.

Store parameters in this order:

KEYSTROKES	OUTPUTS
.739 "A"	0.739
52,941.6 "A"	52,941.600
10.0 "A"	10.000
.75 "A"	.750
5,280 "A"	5,280.000

Compute: Velocity, Reynolds number modified, Reynolds number, critical Reynolds number, type of flow, and pressure drop per mile.

Execution:

Press B	"6.3" ft./sec.
	"84.0"
	"287.47"
	"3.0" (partially turbulent)
	"15.87" PSI/mi.

Optional:

Press D	"1545.89" gpm
	"18.58" BHP/mi.

STEP	KEY ENTRY	KEY CODE
001	f LBL A	31 25 11
	STO (i)	33 24
	f GSB 1	31 22 01
	h RTN	35 22
	f LBL 1	31 25 01
	f ISZ	31 34
	R/S	84
	h ST I	35 33
	GTO 1	22 01
010	h RTN	35 22
	f LBL B	31 25 12
	RCL 0	34 00
	RCL 4	34 04
	x	71
	.	83
	3	03
	2	02
	3	03
	x	71
020	RCL 2	34 02
	÷	81
	STO 5	33 05
	RCL 1	34 01
	.	83
	0	00
	1	01
	1	01
	9	09
	x	71
030	RCL 2	34 02
	g x^2	32 54
	÷	81
	f -x-	31 84
	STO A	33 11
	f GSB 9	31 22 09
	h PAUSE	35 72
	RCL A	34 11
	g x^2	32 54
	RCL D	34 14
040	x	71
	RCL 5	34 05
	x	71
	STO E	33 15
	GTO C	22 13
	f LBL 9	31 25 09
	f GSB 8	31 22 08
	RCL B	34 12
	.	83
	1	01
050	3	03
	5	05
	g $x \leq y$	32 71
	GTO 2	22 02
	h $x \gtrless y$	35 52
	h 1/x	35 62
	.	83

STEP	KEY ENTRY	KEY CODE
	0	00
	0	00
	2	02
060	0	00
	7	07
	x	71
	STO D	33 14
	1	01
	h RTN	35 22
	f LBL 2	31 25 02
	RCL B	34 12
	.	83
	4	04
070	g $x \leq y$	32 71
	GTO 3	22 03
	h $x \gtrless y$	35 52
	.	83
	3	03
	1	01
	9	09
	2	02
	CHS	42
	h y^x	35 63
080	.	83
	0	00
	0	00
	8	08
	1	01
	4	04
	x	71
	STO D	33 14
	2	02
	h RTN	35 22
090	f LBL 3	31 25 03
	RCL B	34 12
	RCL C	34 13
	g $x \leq y$	32 71
	GTO 4	22 04
	h $x \gtrless y$	35 52
	h 1/x	35 62
	.	83
	3	03
	5	05
100	5	05
	h y^x	35 63
	.	83
	0	00
	0	00
	6	06
	6	06
	2	02
	x	71
	.	83
110	0	00
	0	00
	1	01

STEP	KEY ENTRY	KEY CODE
	8	08
	+	61
	STO D	33 14
	3	03
	h RTN	35 22
	f LBL 4	31 25 04
	RCL 6	34 06
120	g x^2	32 54
	h 1/x	35 62
	STO D	33 14
	4	04
	h RTN	35 22
	f LBL 8	31 25 08
	.	83
	0	00
	1	01
	1	01
130	9	09
	RCL 1	34 01
	x	71
	RCL 2	34 02
	RCL 3	34 03
	x	71
	÷	81
	f -x-	31 84
	STO B	33 12
	RCL 2	34 02
140	3	03
	.	83
	7	07
	x	71
	.	83
	0	00
	0	00
	0	00
	7	07
	5	05
150	÷	81
	STO 7	33 07
	1	01
	.	83
	0	00
	5	05
	h y^x	35 63
	1	01
	.	83
	4	04
160	x	71
	RCL 7	34 07
	f LOG	31 53
	4	04
	x	71
	STO 6	33 06
	x	71
	7	07
	7	07

STEP	KEY ENTRY	KEY CODE
	4	04
170	2	02
	÷	81
	STO C	33 13
	f -x-	31 84
	h RTN	35 22
	g LBLf a	32 25 11
	f CL REG	31 43
	h RTN	35 22
	g LBLf b	32 25 12
	.	83
180	7	07
	÷	81
	2	02
	4	04
	x	71
	STO 1	33 01
	h PAUSE	35 72
	h RTN	35 22
	f LBL C	31 25 13
	RCL E	34 15
190	h RTN	35 22
	f LBL D	31 25 14
	RCL 1	34 01
	.	83
	0	00
	2	02
	9	09
	2	02
	x	71
	h PAUSE	35 72
200	RCL E	34 15
	x	71
	1	01
	7	07
	1	01
	5	05
	÷	81
	.	83
	7	07
	7	07
210	÷	81
	h RTN	35 22

LP-D-005 - Liquids Design Parameters Data Card

Card LP-D-005 contains parameters in storage and is defined as a data card.

Storage Register No.	Value
2	.433
6	.0119
7	.241
8	.135
9	.00207
S0	-.3192
S1	.008140
S2	.355
S3	.00662
S4	.0018
S5	USED
S6	1.05
S7	3.7
S8*	.000750
S9	USED

* This register is used for the roughness factor and may be varied by direct input if required.

LP-D-006 - Length Computation

Input:

(1) specific gravity
(2) flowrate - BBLS/Day
(3) internal diameter - inches
(4) viscosity - centistokes
(5) pressure drop - PSI

Computes:

(1) velocity - ft./sec.
(2) Reynolds No. modified - n
(3) Critical Reynolds No. modified - n
(4) type of flow - n
(5) length of line - ft.

Options:

(1) clear for new case
(2) input gpm for flowrate
(3) input new I.D. - inches
(4) input new pressure drop

STEP	INSTRUCTIONS	KEYS	OUTPUT UNITS
1.	load side 1 & 2 of card LP-D-006		
2.	input specific gravity	A	n
3.	input flowrate	A	BBLS/Day
4.	input internal diameter	A	inches
5.	input viscosity	A	centistokes
6.	input pressure drop	A	PSI
7.	Compute:	B	
	(1) velocity		ft./sec.
	(2) Reynolds No. modified		
	(3) Reynolds No. Critical modified		
	(4) type of flow, i.e.		
	1 = viscous		
	2 = intermediate		
	3 = partially turbulent		
	4 = fully turbulent		
	(5) length		feet
8.	Options:		
	(1) clear for new case	fa	
	(2) enter gpm for flowrate	fb	
	(3) enter new I.D.	D	inches
	(4) enter new pressure drop	E	PSI

Note: To compute L, press B after entering option(s) 8 above.

Example LP-D-006

Determine the location of a pump station on a gasoline transmission system with the following parameters:

s.g. of gasoline: .731
flowrate: 22,800 BBLS/Day
internal diameter: 7.981 inches
viscosity: .75 centistokes
pressure at initial station: 850 PSI
pressure at location of new pump station: 32 PSI

Load card LP-D-006, side 1 & 2,
Store parameters in this order:

KEYSTROKES	OUTPUTS
.731 "A"	0.731
22,800 "A"	22,800.000
7.981 "A"	7.981
.75 "A"	.750
818 "A"	818.000

Compute: Velocity, Reynolds No. modified, critical Reynolds No. modified, type of flow, and length of line.

Execution:

Press B	"4.26" ft./sec.
	"45.33" Re*
	"222.12" Re* critical
	"3.0"
	"434,231.0433" ft.
	= 82.24 miles

STEP	KEY ENTRY	KEY CODE
001	f LBL A	31 25 11
	STO (i)	33 24
	f GSB 1	31 22 01
	h RTN	35 22
	f LBL 1	31 25 01
	f ISZ	31 34
	R/S	84
	h ST I	35 33
	GTO 1	22 01
010	h RTN	35 22
	f LBL B	31 25 12
	RCL 4	34 04
	RCL 2	34 02
	x	71
	RCL 0	34 00
	÷	81
	.	83
	3	03
	2	02
020	3	03
	÷	81
	STO 5	33 05
	RCL 1	34 01
	.	83
	0	00
	1	01
	1	01
	9	09
	x	71
030	RCL 2	34 02
	g x^2	32 54
	÷	81
	f -x-	31 84
	STO A	33 11
	f GSB 9	31 22 09
	h PAUSE	35 72
	RCL A	34 11
	g x^2	32 54
	RCL D	34 14
040	x	71
	RCL 5	34 05
	h x≷y	35 52
	÷	81
	STO E	33 15
	GTO C	22 13
	f LBL 9	31 25 09
	f GSB 8	31 22 08
	RCL B	34 12
	.	83
050	1	01
	3	03
	5	05
	g x≤y	32 71
	GTO 2	22 02
	h x≷y	35 52
	h 1/x	35 62

STEP	KEY ENTRY	KEY CODE
	.	83
	0	00
	0	00
060	2	02
	0	00
	7	07
	x	71
	STO D	33 14
	1	01
	h RTN	35 22
	f LBL 2	31 25 02
	RCL B	34 12
	.	83
070	4	04
	g x≤y	32 71
	GTO 3	22 03
	h x≷y	35 52
	.	83
	3	03
	1	01
	9	09
	2	02
	CHS	42
080	h y^x	35 63
	.	83
	0	00
	0	00
	8	08
	1	01
	4	04
	x	71
	STO D	33 14
	2	02
090	h RTN	35 22
	f LBL 3	31 25 03
	RCL B	34 12
	RCL C	34 13
	g x≤y	32 71
	GTO 4	22 04
	h x≷y	35 52
	h 1/x	35 62
	.	83
	3	03
100	5	05
	5	05
	h y^x	35 63
	.	83
	0	00
	0	00
	6	06
	6	06
	2	02
	x	71
110	.	83
	0	00
	0	00

STEP	KEY ENTRY	KEY CODE
	1	01
	8	08
	+	61
	STO D	33 14
	3	03
	h RTN	35 22
	f LBL 4	31 25 04
120	RCL 6	34 06
	g x^2	32 54
	h 1/x	35 62
	STO D	33 14
	4	04
	h RTN	35 22
	f LBL 8	31 25 08
	.	83
	0	00
	1	01
130	1	01
	9	09
	RCL 1	34 01
	x	71
	RCL 2	34 02
	RCL 3	34 03
	x	71
	÷	81
	f -x-	31 84
	STO B	33 12
140	RCL 2	34 02
	3	03
	.	83
	7	07
	x	71
	.	83
	0	00
	0	00
	0	00
	7	07
150	5	05
	÷	81
	STO 7	33 07
	1	01
	.	83
	0	00
	5	05
	h y^x	35 63
	1	01
	.	83
160	4	04
	x	71
	RCL 7	34 07
	f LOG	31 53
	4	04
	x	71
	STO 6	33 06
	x	71
	7	07
	7	07
170	4	04
	2	02
	÷	81
	STO C	33 13
	f -x-	31 84
	h RTN	35 22
	g LBLf a	32 25 11
	f CL REG	31 43
	h RTN	35 22
	g LBLf b	32 25 12
180	.	83
	7	07
	÷	81
	2	02
	4	04
	x	71
	STO 1	33 01
	h RTN	35 22
	f LBL C	31 25 13
	RCL E	34 15
190	h RTN	35 22
	f LBL D	31 25 14
	STO 4	33 04
	h RTN	35 22
	f LBL E	31 25 15
	STO 6	33 06
	h RTN	35 22

LP-D-007 - Internal Diameter Computation

Input:

(1) constant - 0.00
(2) pressure drop per mile - PSI/mi.
(3) .433
(4) flowrate - BBLS/Day
(5) viscosity - centistokes
(6) specific gravity - n

Computes:

(1) internal diameter - inches

Options:

(1) approximate internal diameter for a given velocity

STEP	INSTRUCTIONS	KEYS	OUTPUT UNITS
1.	load card LP-D-005, side 1 & 2		
2.	load card LP-D-007, side 1 & 2		
3.	input constant	A	0.00
4.	input pressure drop per mile	A	PSI/mi.
5.	input .433	A	.433
6.	input viscosity	A	centistokes
7.	input specific gravity	A	n
8.	Compute:		
	(1) internal diameter for a given velocity	D	inches
	(2) internal diameter for actual conditions	E	inches
9.	Option:		
	(1) clear for new case	C	0.00

Example LP-D-007

What is the diameter of a gasoline pipeline 50 miles long when the discharge capacity is 52,941.6 BBLS/Day, the specific gravity is .739, the viscosity is .75 centistokes, and the pressure drop per mile is 15.87 PSI.

Load cards LP-D-005 and LP-D-007, sides 1 & 2. Then store parameters in this order:

KEYSTROKES	OUTPUT
0.00 "A"	0.000
15.87 "A"	15.870
.433 "A"	.433
52,941.6 "A"	52,941.600
.75 "A"	.750
.739 "A"	.739

Compute: Approximate internal diameter based on design velocity; velocity; internal diameter actual conditions.

Execution:

KEYSTROKES	OUTPUTS
Optional: compute approximate I.D. by estimating design velocity:	
7 Press D	9.487
Press E	"6.3"
	"9.996" (Newton Approximation)
Optional: to compute new case:	
Press C	0.00

STEP	KEY ENTRY	KEY CODE
001	f LBL A	31 25 11
	STO (i)	33 24
	f GSB 3	31 22 03
	h RTN	35 22
	f LBL 3	31 25 03
	f ISZ	31 34
	R/S	84
	h ST I	35 33
	GTO 3	22 03
010	h RTN	35 22
	f LBL E	31 25 15
	f FIX	31 23
	GTO B	22 12
	g LBLf a	32 25 11
	RCL B	34 12
	GTO 0	22 00
	f LBL 6	31 25 06
	RCL 0	34 00
	f GSB 1	31 22 01
020	STO B	33 12
	f LBL 0	31 25 00
	RCL A	34 11
	RCL 0	34 00
	STO A	33 11
	-	51
	RCL D	34 14
	RCL B	34 12
	STO D	33 14
	-	51
030	÷	81
	x	71
	STO -0	33 51 00
	RCL 0	34 00
	h PAUSE	35 72
	÷	81
	f RND	31 24
	f x≠0	31 61
	GTO 6	22 06
	g GSB e	32 22 15
040	f -x-	31 84
	RCL 0	34 00
	h RTN	35 22
	f LBL B	31 25 12
	.	83
	0	00
	1	01
	f %	31 82
	STO C	33 13
	2	02
050	÷	81
	-	51
	STO A	33 11
	STO 0	33 00
	f GSB 1	31 22 01
	STO D	33 14
	RCL A	34 11
	RCL C	34 13
	+	61
	STO 0	33 00
060	f GSB 1	31 22 01
	STO B	33 12
	RCL D	34 14
	-	51
	RCL C	34 13
	÷	81
	GTO f a	22 31 11
	f LBL 1	31 25 01
	RCL 9	34 09
	.	83
070	0	00
	0	00
	2	02
	0	00
	7	07
	-	51
	f x=0	31 51
	GTO 2	22 02
	7	07
	ENTER ↑	41
080	0	00
	÷	81
	h RTN	35 22
	f LBL 2	31 25 02
	RCL 7	34 07
	RCL 3	34 03
	g x^2	32 54
	x	71
	RCL 5	34 05
	x	71
090	RCL 1	34 01
	÷	81
	.	83
	2	02
	h y^x	35 63
	f P≷S	31 42
	STO 5	33 05
	f GSB 9	31 22 09
	.	83
	2	02
100	h y^x	35 63
	RCL 5	34 05
	x	71
	f P≷S	31 42
	RCL 0	34 00
	-	51
	h RTN	35 22
	f LBL 9	31 25 09
	f P≷S	31 42
	RCL 6	34 06
110	RCL 3	34 03
	x	71
	RCL 0	34 00

STEP	KEY ENTRY	KEY CODE
	÷	81
	RCL 4	34 04
	÷	81
	f P≷S	31 42
	STO 9	33 09
	.	83
	4	04
120	g x≤y	32 71
	GTO 5	22 05
	RCL 9	34 09
	f P≷S	31 42
	RCL 8	34 08
	g x≤y	32 71
	GTO 7	22 07
	h x≷y	35 52
	h 1/x	35 62
	RCL 9	34 09
130	x	71
	f P≷S	31 42
	h RTN	35 22
	f LBL 5	31 25 05
	RCL 9	34 09
	GTO 8	22 08
	g LBLf d	32 25 14
	g x≤y	32 71
	GTO f c	22 31 13
	h x≷y	35 52
140	h 1/x	35 62
	f P≷S	31 42
	RCL 2	34 02
	h y^x	35 63
	RCL 3	34 03
	x	71
	RCL 4	34 04
	+	61
	h RTN	35 22
	f LBL 8	31 25 08
150	f P≷S	31 42
	RCL 0	34 00
	f P≷S	31 42
	RCL 8	34 08
	÷	81
	RCL 7	34 07
	x	71
	STO 9	33 09
	f LOG	31 53
	4	04
160	x	71
	RCL 9	34 09
	RCL 6	34 06
	h y^x	35 63
	x	71
	1	01
	.	83
	4	04
	x	71
	7	07
170	7	07
	4	04
	2	02
	÷	81
	f P≷S	31 42
	GTO f d	22 31 14
	h RTN	35 22
	f LBL 7	31 25 07
	h x≷y	35 52
	f P≷S	31 42
180	RCL 0	34 00
	h y^x	35 63
	RCL 1	34 01
	x	71
	h RTN	35 22
	g LBLf c	32 25 13
	f P≷S	31 42
	RCL 9	34 09
	f LOG	31 53
	4	04
190	x	71
	g x^2	32 54
	h 1/x	35 62
	f -x-	31 84
	h RTN	35 22
	g LBLf b	32 25 12
	-	51
	RCL 5	34 05
	RCL 2	34 02
	x	71
200	x	71
	RCL 0	34 00
	÷	81
	STO +1	33 61 01
	h RTN	35 22
	g LBLf e	32 25 15
	RCL 6	34 06
	RCL 3	34 03
	x	71
	RCL 0	34 00
210	g x^2	32 54
	÷	81
	h RTN	35 22
	f LBL C	31 25 13
	0	00
	h ST I	35 33
	h RTN	35 22
	f LBL D	31 25 14
	h 1/x	35 62
	RCL 3	34 03
220	RCL 6	34 06
	x	71
	x	71
	f $\sqrt{x}$	31 54
	h RTN	35 22

LP-D-008 - Darcy Weisbach Flowrate Computation

Input:

(1) specific gravity - constant
(2) Δpsi/mi. - constant
(3) I.D. - inches
(4) viscosity - c.s.
(5) internal pipe roughness - x 10^{-6} inch

Computes:

(1) flowrate - BBLS/day using the Darcy Weisbach equation

STEP	INSTRUCTIONS	KEYS	OUTPUT UNITS
1.	load card LP-D-008, side 1 & 2		
2.	input s.g.	STO 1	n
3.	input Δpsi/mi.	STO 2	n
4.	input I.D.	STO 3	inches
5.	input viscosity	STO 4	c.s.
6.	input pipe roughness	STO 5	inches
7.	enter	1.0	1.0
8.	press A	A	1.0
9.	enter initial guess	E	BBLS/day

Example: LP-D-008

Given Conditions:

s.g.: .86815
I.D.: 48"
W.T.: .562"
viscosity: 9.92 c.s.
pipe roughness: 1800 x 10^{-6} inch
P.S. #1 elevation: 37 ft.
P.S. #3 elevation: 1375 ft.
pump suction: 60 psi
length: 104.88 miles
maximum discharge pressure: 1096 psi

$$\Delta\text{psi/mi.} = \frac{(1096 - ((1375 - 37)(.433)(.86815)) - 60)}{104.88 \text{ mi.}}$$

$$= 5.08 \text{ psi/mi.}$$

Execution: Load card LP-D-008, side 1 & 2.

To compute maximum flowrate between P.S. #1 & #3:

Store parameters in this order:

KEYSTROKES	OUTPUT
.86815 "STO 1"	.87
5.08 "STO 2"	5.08
46.876 "STO 3"	46.88

KEYSTROKES	OUTPUT
9.92 "STO 4"	9.9200
1800 x 10^{-6} "STO 5"	.0018
1 "ENTER"	1.0000
"A"	1.0000

Compute Flowrate:
Guess 1,200,000 BPD

1,200,000 "ENTER"___1,200,000

Press E____________1,188,522 BBLS/day - maximum flowrate

STEP	KEY ENTRY	KEY CODE
001	f LBL A	31 25 11
	h ST I	35 33
	h RTN	35 22
	g LBLf e	32 25 15
	h F? 0	35 71 00
	GTO 0	22 00
	h SF 0	35 51 00
	1	01
	h RTN	35 22
010	f LBL 0	31 25 00
	0	00
	h CF 0	35 61 00
	h RTN	35 22
	g LBLf a	32 25 11
	h SF 1	35 51 01
	STO E	33 15
	h RTN	35 22
	f LBL B	31 25 12
	EEX	43
020	CHS	42
	2	02
	RCL E	34 15
	h F? 1	35 71 01
	h x≷y	35 52
	h R↓	35 53
	f %	31 82
	f x=0	31 51
	h LST x	35 82
	STO C	33 13
030	2	02
	÷	81
	-	51
	STO A	33 11
	STO 0	33 00
	f GSB (i)	31 22 24
	STO D	33 14
	RCL A	34 11
	RCL C	34 13
	+	61
040	STO 0	33 00
	f GSB (i)	31 22 24
	STO B	33 12
	RCL D	34 14
	-	51
	RCL C	34 14
	÷	81
	h RTN	35 22
	f LBL C	31 25 13
	STO 0	33 00
050	f GSB (i)	31 22 24
	h RTN	35 22
	f LBL D	31 25 14
	h x≷y	35 52
	STO 0	33 00
	-	51
	h x≷y	35 52
	STO B	33 12
	÷	81
	STO C	33 13
060	2	02
	÷	81
	STO +0	33 61 00
	0	00
	STO 9	33 09
	RCL B	34 12
	h x≷I	35 24
	f LBL 7	31 25 07
	h x≷I	35 24
	STO B	33 12
070	RCL 0	34 00
	f GSB (i)	31 22 24
	RCL C	34 13
	STO +0	33 61 00
	x	71
	STO +9	33 61 09
	RCL B	34 12
	h x≷I	35 24
	f DSZ	31 33
	GTO 7	22 07
080	h ST I	35 33
	RCL 9	34 09
	h RTN	35 22
	f LBL E	31 25 15
	f FIX	31 23
	f GSB B	31 22 12
	RCL B	34 12
	GTO 0	22 00
	f LBL 6	31 25 06
	RCL 0	34 00
090	f GSB (i)	31 22 24
	STO B	33 12
	f LBL 0	31 25 00
	RCL A	34 11
	RCL 0	34 00
	STO A	33 11
	-	51
	RCL D	34 14
	RCL B	34 13
	STO D	33 14
100	-	51
	÷	81
	x	71
	STO -0	33 51 00
	RCL 0	34 00
	h F? 0	35 71 00
	h PAUSE	35 72
	÷	81
	f RND	31 24
	f x≠0	31 61
110	GTO 6	22 06
	RCL 0	34 00
	h RTN	35 22

STEP	KEY ENTRY	KEY CODE
	f LBL 1	31 25 01
	RCL 3	34 03
	2	02
	.	83
	5	05
	h y^x	35 63
	STO 7	33 07
120	RCL 2	34 02
	f $\sqrt{x}$	31 54
	4	04
	.	83
	0	00
	6	06
	4	04
	x	71
	RCL 7	34 07
	x	71
130	STO 7	33 07
	RCL 0	34 00
	÷	81
	RCL 1	34 01
	f $\sqrt{x}$	31 54
	÷	81
	STO 7	33 07
	RCL 0	34 00
	9	09
	2	02
140	.	83
	2	02
	4	04
	x	71
	RCL 3	34 03
	÷	81
	RCL 4	34 04
	÷	81
	RCL 7	34 07
	x	71
150	2	02
	.	83
	5	05
	1	01
	h x⇄y	35 52
	÷	81
	STO 7	33 07
	RCL 5	34 05
	3	03
	.	83
160	7	07
	÷	81
	RCL 3	34 03
	÷	81
	RCL 7	34 07
	+	61
	f LOG	31 53
	2	02
	x	71

STEP	KEY ENTRY	KEY CODE
	CHS	42
170	h 1/x	35 62
	STO 7	33 07
	RCL 3	34 03
	2	02
	.	83
	5	05
	h y^x	35 63
	STO 8	33 08
	RCL 2	34 02
	f $\sqrt{x}$	31 54
180	4	04
	.	83
	0	00
	6	06
	4	04
	x	71
	RCL 8	34 08
	x	71
	RCL 1	34 01
	f $\sqrt{x}$	31 54
190	÷	81
	RCL 7	34 07
	÷	81
	CHS	42
	RCL 0	34 00
	+	61
	h RTN	35 22
	g LBLf c	32 25 13
	STO 1	33 01
	R/S	84
200	STO 2	33 02
	R/S	84
	STO 3	33 03
	R/S	84
	STO 4	33 04
	R/S	84
	STO 5	33 05
	h RTN	35 22

LP-D-009 - Darcy Weisbach Equation Δpsi/mi.

Input:

(1) specific gravity - n
(2) flowrate - BBLS/day
(3) I.D. - inches
(4) viscosity - c.s.
(5) internal pipe roughness - $x10^{-6}$ inch

Computes:

(1) Δpsi/mi.

STEP	INSTRUCTIONS	KEYS	OUTPUT UNITS
1.	load card LP-D-009, side 1 & 2		
2.	input specific gravity	STO 1	n
3.	input flowrate	STO 2	BBLS/day
4.	input I.D.	STO 3	inches
5.	input viscosity	STO 4	c.s.
6.	input pipe roughness	STO 5	inch
7.	enter initial guess of Δpsi/mi.	E	Δ psi/mi.

Example: LP-D-009

Given the conditions of example LP-D-008, compute Δpsi/mi. between P.S. #1 & #3.

Execution: Load card LP-D-009, side 1 & 2.

Store parameters in this order:

KEYSTROKES	OUTPUT
.86815 "STO 1"	.86815
1,188,522 "STO 2"	1,188,522
46.876 "STO 3"	46.876
9.92 "STO 4"	9.92
.00180 "STO 5"	.00180

Compute: Δ psi/mi.

Guess a value - say 10 psi/mi.

Press 10 "E"________5.08 psi/mi.

STEP	KEY ENTRY	KEY CODE
001	f LBL E	31 25 15
	f FIX	31 23
	GTO B	22 12
	g LBLf a	32 25 11
	RCL B	34 12
	GTO 0	22 00
	f LBL 6	31 25 06
	RCL 0	34 00
	f GSB 1	31 22 01
010	STO B	33 12
	f LBL 0	31 25 00
	RCL A	34 11
	RCL 0	34 00
	STO A	33 11
	-	51
	RCL D	34 14
	RCL B	34 12
	STO D	33 14
	-	51
020	÷	81
	x	71
	STO -0	33 51 00
	RCL 0	34 00
	h PAUSE	35 72
	÷	81
	f RND	31 24
	f x≠0	31 61
	GTO 6	22 06
	RCL 0	34 00
030	h RTN	35 22
	f LBL B	31 25 12
	.	83
	0	00
	1	01
	f %	31 82
	STO C	33 13
	2	02
	÷	81
	-	51
040	STO A	33 11
	STO 0	33 00
	f GSB 1	31 22 01
	STO D	33 14
	RCL A	34 11
	RCL C	34 13
	+	61
	STO 0	33 00
	f GSB 1	31 22 01
	STO B	33 12
050	RCL D	34 14
	-	51
	RCL C	34 13
	÷	81
	GTO f a	22 31 11
	f LBL 1	31 25 01
	RCL 3	34 03
	2	02
	.	83
	5	05
060	h y^x	35 63
	RCL 0	34 00
	f $\sqrt{x}$	31 54
	x	71
	4	04
	.	83
	0	00
	6	06
	4	04
	x	71
070	RCL 2	34 02
	÷	81
	RCL 1	34 01
	f $\sqrt{x}$	31 54
	÷	81
	STO 7	33 07
	RCL 2	34 02
	x	71
	9	09
	2	02
080	.	83
	2	02
	4	04
	x	71
	RCL 3	34 03
	÷	81
	RCL 4	34 04
	÷	81
	2	02
	.	83
090	5	05
	1	01
	h x≷y	35 52
	÷	81
	STO 7	33 07
	RCL 5	34 05
	3	03
	.	83
	7	07
	÷	81
100	RCL 3	34 03
	÷	81
	RCL 7	34 07
	+	61
	STO 7	33 07
	f LOG	31 53
	2	02
	x	71
	CHS	42
	h 1/x	35 62
110	STO 7	33 07
	RCL 3	34 03
	2	02

STEP	KEY ENTRY	KEY CODE
	.	83
	5	05
	h y^x	35 63
	RCL 0	34 00
	f $\sqrt{x}$	31 54
	x	71
	4	04
120	.	83
	0	00
	6	06
	4	04
	x	71
	RCL 1	34 01
	f $\sqrt{x}$	31 54
	÷	81
	RCL 7	34 07
	÷	81
130	CHS	42
	RCL 2	34 02
	+	61
	h RTN	35 22

LP-D-010 - Darcy Weisbach Hydraulic Gradient Computation

Input:

(1) specific gravity - n
(2) flowrate - BBLS/day
(3) I.D. - inches
(4) viscosity - c.s.
(5) internal pipe roughness - inch
(6) pumping station elevation - ft.
(7) receiving station elevation - ft.
(8) distance between pumping and receiving stations - mile
(9) suction pressure at receiving station - psi

Computes:

(1) Δpsi due to elevation difference between pumping station and receiving station - psi.
(2) Δpsi/mi. at given flowrate and 0 ft. elevation difference between pumping station and receiving station - psi/mi.
(3) total psi for length of segment - psi
(4) total psi due to elevation - psi
(5) total discharge pressure at pumping station with 0 psi suction at receiving station - psi
(6) total discharge pressure at pumping station and given suction condition at receiving station - psi
(7) same as item (6) above in feet of head - ft.
(8) total discharge pressure at pumping station in feet of head plus the pumping station elevation - ft.
(9) Options:
    (a) repeat procedure from item (1) thru (8)
    (b) compute differential BHP required at pumping station based on .77 efficiency.

STEP	INSTRUCTIONS	KEYS	OUTPUT UNITS
1.	load card LP-D-010, side 1 & 2		
2.	input specific gravity	STO 1	n
3.	input flowrate	STO 2	BBLS/day
4.	input pipeline I.D.	STO 3	inches
5.	input viscosity	STO 4	c.s.
6.	input pipe internal roughness	STO 5	inch
7.	input pumping station elevation	STO 6	ft.
8.	input receiving station elevation	STO 7	ft.
9.	input distance between pumping station & receiving station	STO 8	mi.
10.	input suction pressure at receiving station	STO 9	psi
11.	compute Δpsi due to elevation differences	C	psi
12.	guess Δpsi/mi. and compute Δpsi/mi.	E	psi/mi.
	(a) compute total friction psi		psi
	(b) compute total differential elevation		psi
	(c) option to add or subtract		+ or -

(Uphill flow between pumping & receiving station = +)
(Downhill flow between pumping & receiving station = -)

STEP	INSTRUCTIONS	KEYS	OUTPUT UNITS
13.	To continue execution	R/S	
	(a) computes total discharge pressure plus suction		psi
	(b) computes total discharge pressure plus suction		ft.
	(c) computes total discharge pressure plus suction plus pumping station elevation		ft.
14.	Option to go to step 12(a) and repeat computation	fe	
15.	Option to compute differential BHP @ .77 efficiency	D	BHP

Example: LP-D-010

Given Conditions:

s.g.: .858
Q: 1,200,000 BBLS/day
I.D.: 46.876 inches
ν: 9.11 c.s.
pipe roughness: 1800 x 10^{-6} inch
pumping station elevation: 37'
receiving station elevation: 1375'
length: 104.88 miles
suction pressure at receiving point: 60 psi

Compute the hydraulic gradient parameters at the pumping station.

Execution: Load card LP-D-010, side 1 & 2

Store parameters in this order:

KEYSTROKES	OUTPUT
.858 "STO 1"	.858
1,200,000 "STO 2"	1,200,000
46.876 "STO 3"	46.876
9.11 "STO 4"	9.11
.0018 "STO 5"	.0018
37 "STO 6"	37
1375 "STO 7"	1375
104.88 "STO 8"	104.88
60 "STO 9"	60

To Compute Hydraulic Gradient Parameters:

Press C	497.09 psi
Guess "5" ENTER	5.00
Press E	"5.03" psi/mi.
	"527.98" psi - friction
	"497.09" psi - elevation

Press +__________ 1025 - total discharge pressure less suction at receiving station - psi

Press R/S__________ "1085.1" - total discharge pressure plus suction

__________ "2920.7" - total discharge pressure plus suction - ft.

__________ "2957.7" - total discharge pressure plus suction plus station elevation

Option: Press D__________ 28,759 BHP @ .77 efficiency

RCL A__________ 5.03 psi/mi.

f P≷S "RCL 0"__________ 497.09 psi - elevation

f P≷S "RCL 2"__________ 2957.7 ft. - total head

fe__________ return program to step 12 (a)

STEP	KEY ENTRY	KEY CODE
001	f LBL E	31 25 15
	f FIX	31 23
	GTO B	22 12
	g LBLfa	32 25 11
	RCL B	34 12
	GTO 0	22 00
	f LBL 6	31 25 06
	RCL 0	34 00
	f GSB 1	31 22 01
010	STO B	33 12
	f LBL 0	31 25 00
	RCL A	34 11
	RCL 0	34 00
	STO A	33 11
	-	51
	RCL D	34 14
	RCL B	34 12
	STO D	33 14
	-	51
020	÷	81
	x	71
	STO -0	33 51 00
	RCL 0	34 00
	h PAUSE	35 72
	÷	81
	f RND	31 24
	f x≠0	31 61
	GTO 6	22 06
	RCL 0	34 00
030	h PAUSE	35 72
	f P≷S	31 42
	STO 3	33 03
	f P≷S	31 42
	g LBL e	32 25 15
	f P≷S	31 42
	RCL 3	34 03
	f P≷S	31 42
	RCL 8	34 08
	x	71
040	h PAUSE	35 72
	f P≷S	31 42
	RCL 0	34 00
	f P≷S	31 42
	R/S	84
	+	61
	RCL 9	34 09
	+	61
	h PAUSE	35 72
	.	83
050	4	04
	3	03
	3	03
	÷	81
	RCL 1	34 01
	÷	81
	h PAUSE	35 72
	RCL 6	34 06
	+	61
	f P≷S	31 42
060	STO 2	33 02
	RCL 1	34 01
	f P≷S	31 42
	STO 7	33 07
	f P≷S	31 42
	RCL 2	34 02
	f P≷S	31 42
	h RTN	35 22
	f LBL B	31 25 12
	.	83
070	0	00
	1	01
	f %	31 82
	STO C	33 13
	2	02
	÷	81
	-	51
	STO A	33 11
	STO 0	33 00
	f GSB 1	31 22 01
080	STO D	33 14
	RCL A	34 11
	RCL C	34 13
	+	61
	STO 0	33 00
	f GSB 1	31 22 01
	STO B	33 12
	RCL D	34 14
	-	51
	RCL C	34 13
090	÷	81
	GTO f a	22 31 01
	f LBL 1	31 25 01
	RCL 3	34 03
	2	02
	.	83
	5	05
	h y^x	35 63
	RCL 0	34 00
	f $\sqrt{x}$	31 54
100	x	71
	4	04
	.	83
	0	00
	6	06
	4	04
	x	71
	RCL 2	34 02
	÷	81
	RCL 1	34 01
110	f $\sqrt{x}$	31 54
	÷	81
	STO 7	33 07

STEP	KEY ENTRY	KEY CODE
	RCL 2	34 02
	x	71
	9	09
	2	02
	.	83
	2	02
	4	04
120	x	71
	RCL 3	34 03
	÷	81
	RCL 4	34 04
	÷	81
	2	02
	.	83
	5	05
	1	01
	h x≷y	35 52
130	÷	81
	STO 7	33 07
	RCL 5	34 05
	3	03
	.	83
	7	07
	÷	81
	RCL 3	34 03
	÷	81
	RCL 7	34 07
140	+	61
	STO 7	33 07
	f LOG	31 53
	2	02
	x	71
	CHS	42
	h 1/x	35 62
	STO 7	33 07
	RCL 3	34 03
	2	02
150	.	83
	5	05
	h y^x	35 63
	RCL 0	34 00
	f $\sqrt{x}$	31 54
	x	71
	4	04
	.	83
	0	00
	6	06
160	4	04
	x	71
	RCL 1	34 01
	f $\sqrt{x}$	31 54
	÷	81
	RCL 7	34 07
	÷	81
	CHS	42
	RCL 2	34 02

STEP	KEY ENTRY	KEY CODE
	+	61
170	h RTN	35 22
	f LBL C	31 25 13
	RCL 6	34 06
	RCL 7	34 07
	f P≷S	31 42
	STO 1	33 01
	f P≷S	31 42
	-	51
	h ABS	35 64
	.	83
180	4	04
	3	03
	3	03
	x	71
	RCL 1	34 01
	x	71
	f P≷S	31 42
	STO 0	33 00
	f P≷S	31 42
	h RTN	35 22
190	f LBL D	31 25 14
	RCL 6	34 06
	f P≷S	31 42
	RCL 2	34 02
	f P≷S	31 42
	-	51
	h ABS	35 64
	.	83
	4	04
	3	03
200	3	03
	x	71
	RCL 1	34 01
	x	71
	RCL 2	34 02
	x	71
	2	02
	4	04
	÷	81
	2	02
210	4	04
	5	05
	0	00
	÷	81
	.	83
	7	07
	7	07
	÷	81
	h RTN	35 22

LP-D-011 - Modified Darcy Weisbach Computation - Large Diameter Lines

Input:

(1) I.D. - inches
(2) viscosity - c.s.
(3) Δpsi/mi.
(4) specific gravity - n
(5) internal pipe roughness - x 10^{-6} inch

Computes:

(1) flowrate - BBLS/day

STEP	INSTRUCTIONS	KEYS	OUTPUT UNITS
1.	load card LP-D-011, side 1 & 2		
2.	input I.D.	STO 1	inches
3.	input viscosity	STO 2	c.s.
4.	input psi/mi.	STO 3	psi
5.	input specific gravity	STO 4	n
6.	input internal pipe roughness	STO 5	n inch
7.	compute flowrate, guess n BBLS/day	E	BBLS/day

Example: LP-D-011

Given Conditions:

I.D.: 46.876 inches
viscosity: 9.11 c.s.
Δpsi/mi.: 5.14
s.g.: .858
e: .0007 inch

Compute flowrate

Execution: Load card LP-D-011, side 1 & 2

Store parameters in this order:

KEYSTROKES	OUTPUT
46.876 "STO 1"	46.8760
9.11 "STO 2"	9.1100
5.14 "STO 3"	5.1400
.858 "STO 4"	.8580
.0007 "STO 5"	.0007

Guess value & Compute flowrate:

1,200,000 "E"________1,096,000 BBLS/day

STEP	KEY ENTRY	KEY CODE
001	f LBL E	31 25 15
	f FIX	31 23
	GTO B	22 12
	g LBLf a	32 25 11
	RCL B	34 12
	GTO 0	22 00
	f LBL 6	31 25 06
	RCL 0	34 00
	f GSB 1	31 22 01
010	STO B	33 12
	f LBL 0	31 25 00
	RCL A	34 11
	RCL 0	34 00
	STO A	33 11
	-	51
	RCL D	34 14
	RCL B	34 12
	STO D	33 14
	-	51
020	÷	81
	x	71
	STO -0	33 51 00
	RCL 0	34 00
	h PAUSE	35 72
	÷	81
	f RND	31 24
	f x≠0	31 61
	GTO 6	22 06
	RCL 0	34 00
030	h RTN	35 22
	f LBL B	31 25 12
	.	83
	0	00
	1	01
	f %	31 82
	STO C	33 13
	2	02
	÷	81
	-	51
040	STO A	33 11
	STO 0	33 00
	f GSB 1	31 22 02
	STO D	33 14
	RCL A	34 11
	RCL C	34 13
	+	61
	STO 0	33 00
	f GSB 1	31 22 01
	STO B	33 12
050	RCL D	34 14
	-	51
	RCL C	34 13
	÷	81
	GTO f a	22 31 11
	f LBL 1	31 25 01
	RCL 3	34 03
	RCL 4	34 04
	÷	81
	f $\sqrt{x}$	31 54
060	RCL 1	34 01
	2	02
	.	83
	5	05
	h y^x	35 63
	4	04
	.	83
	0	00
	6	06
	4	04
070	x	71
	RCL 0	34 00
	h x≷y	35 52
	÷	81
	2	02
	x	71
	1	01
	.	83
	4	04
	1	01
080	2	02
	5	05
	x	71
	STO 7	33 07
	RCL 0	34 00
	9	09
	2	02
	.	83
	2	02
	4	04
090	x	71
	RCL 1	34 01
	÷	81
	RCL 2	34 02
	÷	81
	STO 8	33 08
	RCL 7	34 07
	h x≷y	35 52
	÷	81
	STO 9	33 09
100	RCL 5	34 05
	3	03
	.	83
	7	07
	÷	81
	RCL 1	34 01
	÷	81
	RCL 9	34 09
	+	61
	f LOG	31 53
110	2	02
	CHS	42
	x	71

STEP	KEY ENTRY	KEY CODE
	STO 9	33 09
	RCL 3	34 03
	RCL 4	34 04
	÷	81
	f $\sqrt{x}$	31 54
	RCL 1	34 01
	2	02
120	.	83
	5	05
	h y^x	35 63
	x	71
	4	04
	.	83
	0	00
	6	06
	4	04
	x	71
130	RCL 9	34 09
	x	71
	RCL 0	34 00
	CHS	42
	+	61
	h RTN	35 22

References

1-1 T.R. Aude, "Suggested Formula for Calculating Capacity of Products Pipe Lines", 1943.

1-2 C.F. Colebrook, "Turbulent Flow in Pipes with Particular Reference to the Transition Region Between the Smooth and Rough Pipe Laws", J. Inst. Civil Engineers, London, 1939.

1-3 H.D'Arcy, "Recherches Experimentales Relatives au Mouvement d'Eau dans les Tuyaux", Mallet-Bachelier, Paris, 1857.

1-4 W.G. Heltzel, "Fluid Flow and Friction in Pipelines". Oil & Gas Journal, June 5, 1930.

1-5 R.W. Leach & W.P. Redmond, "How a Computer is Applied to a Specific Problem of Pipeline Design". Oil & Gas Journal, Flow Calculations in Pipelining, 1960.

1-6 L.F. Moody, "Friction Factors for Pipe Flow:, Trans Am. Society Mechanical Engineers, 1944.

1-7 Wilson & McAdams, "Contribution No. 19" from the Department of Chemical Engineering, Massachusetts Institute of Technology.

1-8 Wilson, McAdams & Seltzer, "Flow of Fluids Through Commercial Pipelines", Journal of Industrial and Engineering Chemistry, February, 1922.

1-9 T.R. Young, "Digital Simulation of Crude Oil Pipelines", API Pipeline Report, May 2, 1960.

1-10 Technical Paper No. 410, "Flow of Fluids Through Valves, Fittings and Pipe", Crane Company, 1969.

1-11 NGPSA Engineering Data Book, 1972.

1-12 "Steady Flow in Gas Pipelines", American Gas Association, 1965.

Nomenclature

A - $\frac{(D^2 - d^2)\pi}{4}$ = cross sectional metal area, sq. in.

A_f - $\frac{d^2\pi}{4}$ = cross sectional flow area, sq. in.

A_i - $\frac{d\pi}{12}$ = inside pipe surface, sq. ft. per ft. length

A_o - $\frac{D\pi}{12}$ = outside pipe surface, sq. ft. per ft. length

d - inside diameter of pipe in inches

D - outside diameter of pipe, inches

E - longitudinal joint factor. Normally a factor of 1.0 is used for seamless and welded pipe except for the following:

Fusion Welded A 134 and A 139	0.80
Spiral Welded A 211	0.80
Butt Welded ASTM-A53, API-5L	0.60

F_f - drag factor

f - friction factor

f_i - intermediate flow friction factor

f_{ft} - fully turbulent flow friction factor

f_{pt} - partially turbulent flow friction factor

F - construction type design factor

f_v - viscous flow friction factor

I - Ar_g = $0.0491(D^4 - d^4)$ = moment of inertia, in.^{4}

k_e - operating roughness factor in micro inches

L - length of line in feet

n - a constant

p - design pressure, psig

ΔP - pressure drop in PSI

Q - quantity flow in barrels per day

R^* - Reynolds number, modified ($Re \div 7742$)

r_g - $\sqrt{I/A}$ = $\frac{\sqrt{D^2 + d^2}}{4}$ = radius of gyration, inches

S - specified minimum yield strength, PSI

s.g. - specific gravity

T - temperature derating factor

t - nominal wall thickness, inches

U - Saybolt universal viscosity, seconds

v - average velocity in feet per second

w - $3.4A$ = weight of pipe, lb. per ft. length

w_w - $0.433A_f$ = weight of water filling, lb. per ft. length

Z - $\frac{2I}{D}$ = $0.0982\,\frac{(D^4 - d^4)}{D}$ = section modulus, in.^{3}

z - absolute viscosity in centipoises

z - kinematic viscosity

2.0 Gas Pipeline Design

2.1 Introduction

The Gas Pipeline Design series contains five basic programs for the detailed hydraulic design of gas gathering/transmission systems.

2.2 Theoretical Formulae

The Gas Design programs are all based on the Institute of Gas Technology technical report No. 10, "Steady Flow in Gas Pipelines".

The following information is quoted from this report.

Gas Hydraulics Formulae

$$Q_b = 38.774 \frac{T_b}{P_b} \sqrt{1/f} \left[\frac{P_1^2 - P_2^2 \quad -.0375 \dfrac{G \Delta H \, P_{avg}^2}{Tz}}{GLTz} \right]^{.5} D^{2.5}$$

$$\sqrt{1/f_{pt}} = 4 F_f \log (N_{re} / 1.4 \sqrt{1/f_{pt}}$$

$$\sqrt{1/f_{ft}} = 4 \log (3.7 \, D/k_e)$$

$$N_{re} \text{ (critical)} = 1.4 \, (3.7D/k_e)^{1/F_f} \sqrt{1/f_{ft}}$$

$$F_f = \exp \left[\frac{BI + 87.3}{-3593} \right] \qquad BI = (°\text{Bend/mile})$$

$$N_{re} = .4775 \frac{Q_b \, G}{\mu \text{ I.D.}} \frac{(P_b)}{(T_b)}$$

$$P_m = .67 \left[\frac{(P_1^3 - P_2^3)}{(P_1^2 - P_2^2)} \right]$$

$$P_1 = \left[\left[\frac{25.79 \, (Q_b \text{ mcf}) (P_b)}{\sqrt{1/f} \, (T_b) (D^{2.5})} \right]^2 GLTz + P_2^2 + \Delta H \right]^{.5}$$

$$P_2 = \left[\left[P_1^2 - \frac{(25.79) (Q_b \text{mcf}) (P_b)}{(\sqrt{1/f}) (T_b) (D^{2.5})} \right]^2 GLTz - \Delta H \right]^{.5}$$

$$L = \left[\frac{(P_1^2 - P_2^2 - \Delta H)}{GTz} \right] \left[\frac{(25.79) (Q_b) (P_b)}{(\sqrt{1/f}) (T_b) (D^{2.5})} \right]^{-2}$$

2.3 Determination of Applicable Equation

The basic programs utilize a Newton Approximation technique

coupled with a Reynolds number comparison to select the proper transmission factor. Flow above the critical Reynolds number is fully turbulent, below is partially turbulent.

2.4 Pressure Drop Computation

The gas pressure drop computation is obtained by using the P_1 or P_2 programs and solving for the appropriate upstream or downstream pressure, ie., given P_1 solve for P_2 and divide by the distance: $\Delta PSI/L$.

2.5 Compressibility Factor

The compressibility factors are obtained by using the following formulae:

(a) critical pressure: $696.5 - 43G$ (psia)

(b) critical temperature: $180 + 300G$ (°R)

(c) reduced pressure: $P_m/P_c = P_r$

(d) reduced temperature: $T_f/T_c = T_r$

(e) $z = 1 + (-.60410 + .34098\ T_r)\ P_r$

This equation is valid only for $0 \leq P_r \leq 2$
$1.25 \leq T_r \leq 1.6$

2.6 Operating Roughness

Bare Steel Pipe	Roughness (micro inches)
new	500 - 700
6 months	1000 - 1250
12 months	1500
24 months	1750
plastic lined	200 - 300
sand blasted	200 - 300
pig burnished	300 - 500
aluminum	200

2.7 Weymouth Approximation

$$Q = 871 \times d^{2.667} \times \sqrt{P_1^2 - P_2^2} \times \frac{1}{\sqrt{L}}$$, where

Q = cfd
L = miles

P_1 = psia , P_2 = psia

d = I.D., inches

$$d = \left[\frac{Q \times L^{1/2}}{871 \times (P_1^2 - P_2^2)^{1/2}}\right]^{.375}$$

$$L = \left[\frac{871}{Q} \times d^{2.667} \times (P_1^2 - P_2^2)^{1/2}\right]^2$$

$$P_1 = \left[\left[\frac{Q \times L^{1/2}}{871 \times d^{2.667}}\right]^2 + P_2^2\right]^{1/2}$$

$$P_2 = \left[(P_1^2) - \left[\frac{Q \times L^{1/2}}{871 \times d^{2.667}}\right]^2\right]^{1/2}$$

GP-D-001A - Volumetric Flowrate Computations
&
GP-D-001B

Input:

(1) outside diameter - inches
(2) wall thickness - inches
(3) inlet pressure - psig
(4) outlet pressure - psig
(5) barometric pressure - psia
(6) specific gravity - n
(7) length of the pipeline - miles
(8) temperature of the flowing gas - ^oF
(9) base pressure - psia
(10) base temperature - ^oF
(11) viscosity - lb.m./ft.sec. - n
(12) roughness factor - micro inches
(13) drag factor - n
(14) elevation difference - feet

Computes:

(1) average line pressure - psig
(2) critical pressure - psia
(3) critical temperature - ^oR
(4) reduced pressure - n
(5) reduced temperature - n
(6) compressibility - n
(7) α factor - n
(8) ψ factor - n
(9) Weymouth approximation - mcfd
(10) Newton secant approximation - mcfd
(11) Reynolds no. - n
(12) critical Reynolds no. - n
(13) transmission factor $\sqrt{1/f_{ft}}$ - n
(14) transmission factor $\sqrt{1/f_{pt}}$ - n

STEP	INSTRUCTIONS	KEYS	OUTPUT UNITS
1.	load card GP-D-001A, side 1 & 2		
2.	input O.D.	A	inches
3.	input W.T.	A	inches
4.	input inlet pressure	A	psig
5.	input outlet pressure	A	psig
6.	input barometric pressure	A	psia
7.	input specific gravity	A	n
8.	input length of pipeline	A	miles
9.	input temperature of flowing gas	A	^oF
10.	input base pressure	A	psia
11.	input base temperature	A	^oF
12.	input viscosity	A	lb.m./ft.sec.
13.	input roughness factor	A	micro inches
14.	input drag factor	A	n
15.	input elevation difference	A	feet

STEP	INSTRUCTIONS	KEY	OUTPUT UNITS
16.	Compute:	B	
	(1) average base pressure		psig
	(2) critical pressure		psia
	(3) critical temperature		^oR
	(4) reduced pressure		n
	(5) reduced temperature		n
	(6) compressibility		n
	(7) α factor		n
17.	load card GP-D-001B side 1 & 2		
18.	Compute throughput:		
	(1) Weymouth approximation	A	mcfd
	(2) (a) Newton secant approximation	E	mcfd
	(b) Reynolds no.		n
	(c) critical Reynolds no.		n
	(d) transmission factor $\sqrt{1/f_{ft}}$		n
	(e) transmission factor $\sqrt{1/f_{pt}}$		n
	(f) throughput		mcfd
19.	Do you want to run a new case: If yes: (1) f CL REG		
	(2) CL x		0.00

Example: GP-D-001A & B

The following information is given for a gas transmission pipeline. Determine the flowrate in mcfd.

(1) Calculation of Line Throughput - Fully Turbulent Flow

Given Conditions:

outside pipe diameter: 20 inches
wall thickness: .281 inches
upstream pressure: 586 psig
downstream pressure: 486 psig
barometric pressure: 14 psia
specific gravity of gas: .570
line length: 10 miles
flowing temperature: 70^oF
base pressure: 14.65 psia
base temperature: 50^oF
viscosity: 7.95×10^{-6} lb.m./ft.sec.
effective roughness: 1250 micro inches
drag factor: .9545
elevation difference: 0

Execution: Load card GP-D-001A, side 1 & 2

Store parameters in this order:

KEYSTROKES	OUTPUT
20 "A"	20.0000
.281 "A"	.2810
586 "A"	586.0000
486 "A"	486.0000
14 "A"	14.0000
.570 "A"	.5700
10 "A"	10.0000
70 "A"	70.0000
14.65 "A"	14.6500
50 "A"	50.0000
7.95×10^{-6} "A"	7.95×10^{-6}
1250 "A"	1,250.0000
.9545 "A"	.9545
0 "A"	0.0000

To Compute:

Press "B"__________"540.27" psig (average pressure)
__________"671.99" psia (critical pressure)
__________"351.00" °R (critical temperature)
__________".82" (reduced pressure)
__________"1.51" (reduced temperature)
__________".93" (compressibility)
__________"50.59" (α coefficient)

Load card GP-D-001B_50.59376

Press "A"__________248,779.88 mcfd (Weymouth Approx.)

Press "E"__________268,061.54 mcfd (Newton Approx.)

Press "C"__________"13,562,244.27" (Reynolds No.)
__________"2,586,035.79" (critical Reynolds No.)
__________"19.03977" (Transmission factor for fully turbulent condition)
__________".00001" (verifies fully turbulent region Re No. > critical Reynolds No.)
__________"268,061.54" mcfd (recalls flowrate for actual conditions)

New Case Desired?
f CL REG
CL x__________0.00
Proceed to step 1

(2) Calculation of Line Throughput - Partially Turbulent Flow

Given Conditions:

outside pipe diameter: 20 inches
wall thickness: .281 inches
upstream pressure: 585 psig
downstream pressure: 555 psig
barometric pressure: 15 psia
specific gravity of gas: .570
line length: 20 miles
flowing temperature: 70°F
base pressure: 14.65 psia
base temperature: 50°F
viscosity: 7.95 x 10^{-6} lb.m./ft.sec.
effective roughness: 300 micro inches
drag factor: .9720
elevation difference: 0

Execution: Load card GP-D-001A, side 1 & 2

Store parameters in this order:

KEYSTROKES	OUTPUT
20 "A"	20.0000
.281 "A"	.2810
585 "A"	585.0000
555 "A"	555.0000
15 "A"	15.0000
.570 "A"	.5700
20 "A"	20.0000
70 "A"	70.0000
14.65 "A"	14.6500
50 "A"	50.0000
7.95 x 10^{-6} "A"	7.95 x 10^{-6}
300 "A"	300.0000
.9720 "A"	.9720
0 "A"	0.0000

To Compute:

Press "B"	"573.05" psig (average pressure)
	"671.99" psia (critical pressure)
	"351.00" °R (critical temperature)
	".88" (reduced pressure)
	"1.51" (reduced temperature)
	".92" (compressibility)
	"50.59" (α coefficient)

Load card GP-D-001B

Press "A"	99,360.99 mcfd (Weymouth Approx.)
Press "E"	116,352.27 mcfd (Newton Approx.)
Press "C"	"5,886,672.945" (Reynolds No.)
	"10,319,351.89" (critical Reynolds No.)

KEYSTROKES	OUTPUT
"C"	"21.51893" (Transmission factor for fully turbulent condition)
	"20.64140" (Transmission factor for partially turbulent condition)
	"116,352.2732" mcfd (Newton Approx. volumetric flowrate)

New Case Desired?

f CL REG

CL x ________ 0.00

Proceed to step 1

STEP	KEY ENTRY	KEY CODE
001	f LBL A	31 25 11
	STO (i)	33 24
	f GSB 1	31 22 01
	h RTN	35 22
	f LBL 1	31 25 01
	f ISZ	31 34
	R/S	84
	h ST I	35 33
	GTO 1	22 01
010	h RTN	35 22
	f LBL B	31 25 12
	RCL 2	34 02
	RCL 4	34 04
	+	61
	3	03
	h y^x	35 63
	RCL 3	34 03
	RCL 4	34 04
	+	61
020	3	03
	h y^x	35 63
	-	51
	STO (i)	33 24
	RCL 2	34 02
	RCL 4	34 04
	+	61
	g x^2	32 54
	RCL 3	34 03
	RCL 4	34 04
030	+	61
	g x^2	32 54
	-	51
	RCL (i)	34 24
	h $x \gtrless y$	35 52
	STO (i)	33 24
	÷	81
	.	83
	6	06
	7	07
040	x	71
	RCL 4	34 04
	-	51
	f -x-	31 84
	STO A	33 11
	RCL 5	34 05
	4	04
	3	03
	x	71
	CHS	42
050	6	06
	9	09
	6	06
	.	83
	5	05
	+	61
	STO B	33 12

STEP	KEY ENTRY	KEY CODE
	f -x-	31 84
	RCL 5	34 05
	3	03
060	0	00
	0	00
	x	71
	1	01
	8	08
	0	00
	+	61
	STO C	33 13
	f -x-	31 84
	RCL A	34 11
070	RCL 4	34 04
	+	61
	RCL B	34 12
	÷	81
	f -x-	31 84
	STO D	33 14
	RCL 7	34 07
	4	04
	6	06
	0	00
080	+	61
	RCL C	34 13
	÷	81
	f -x-	31 84
	STO E	33 15
	1	01
	.	83
	2	02
	5	05
	g $x \le y$	32 71
090	GTO 2	22 02
	RCL E	34 15
	.	83
	9	09
	x	71
	1	01
	.	83
	3	03
	-	51
	RCL D	34 14
100	x	71
	1	01
	+	61
	f -x-	31 84
	STO B	33 12
	GTO 3	22 03
	f LBL 2	31 25 02
	RCL E	34 15
	.	83
	3	03
110	4	04
	x	71
	.	83

STEP	KEY ENTRY	KEY CODE
	6	06
	-	51
	RCL D	34 14
	x	71
	1	01
	+	61
	STO B	33 12
120	f -x-	31 84
	GTO 3	22 03
	f LBL 3	31 25 03
	RCL (i)	34 24
	f GSB 4	31 22 04
	-	51
	RCL 5	34 05
	÷	81
	RCL 6	34 06
	÷	81
130	RCL 7	34 07
	4	04
	6	06
	0	00
	+	61
	÷	81
	RCL B	34 12
	÷	81
	f$\sqrt{x}$	31 54
	GTO 5	22 05
140	f LBL 4	31 25 04
	RCL A	34 11
	g x^2	32 54
	RCL 5	34 05
	x	71
	f P≷S	31 42
	RCL 3	34 03
	x	71
	f P≷S	31 42
	RCL 7	34 07
150	4	04
	6	06
	0	00
	+	61
	÷	81
	RCL B	34 12
	÷	81
	.	83
	0	00
	3	03
160	7	07
	5	05
	x	71
	h RTN	35 22
	f LBL 5	31 25 05
	ENTER ↑	41
	RCL 0	34 00
	RCL 1	34 01
	2	02
	x	71
170	-	51
	f P≷S	31 42
	STO 5	33 05
	f P≷S	31 42
	2	02
	.	83
	5	05
	h y^x	35 63
	x	71
	3	03
180	8	08
	.	83
	7	07
	7	07
	x	71
	RCL 9	34 09
	4	04
	6	06
	0	00
	+	61
190	RCL 8	34 08
	÷	81
	x	71
	1	01
	0	00
	0	00
	0	00
	÷	81
	f P≷S	31 42
	STO 8	33 08
200	RCL 0	34 00
	RCL 5	34 05
	x	71
	f P≷S	31 42
	RCL 9	34 09
	4	04
	6	06
	0	00
	+	61
	x	71
210	h 1/x	35 62
	RCL 5	34 05
	x	71
	RCL 8	34 08
	x	71
	.	83
	4	04
	7	07
	7	07
	5	05
220	x	71
	f P≷S	31 42
	STO 9	33 09
	f P≷S	31 42
	h RTN	35 22

STEP	KEY ENTRY	KEY CODE
001	f LBL E	31 25 15
	f FIX	31 23
	GTO B	22 12
	g LBLf a	32 25 11
	RCL 2	34 02
	GTO 0	22 00
	f LBL 6	31 25 06
	RCL 0	34 00
	f GSB 9	31 22 09
010	STO 2	33 02
	f LBL 0	31 25 00
	RCL 1	34 01
	RCL 0	34 00
	STO 1	33 01
	-	51
	RCL 4	34 04
	RCL 2	34 02
	STO 4	33 04
	-	51
020	÷	81
	x	71
	STO -0	33 51 00
	RCL 0	34 00
	h PAUSE	35 72
	÷	81
	f RND	31 24
	f x≠0	31 61
	GTO 6	22 06
	RCL 0	34 00
030	h RTN	35 22
	f LBL B	31 25 12
	.	83
	0	00
	1	01
	f %	31 82
	STO 3	33 03
	2	02
	÷	81
	-	51
040	STO 1	33 01
	STO 0	33 00
	f GSB 9	31 22 09
	STO 4	33 04
	RCL 1	34 01
	RCL 3	34 03
	+	61
	STO 0	33 00
	f GSB 9	31 22 09
	STO 2	33 02
050	RCL 4	34 04
	-	51
	RCL 3	34 03
	÷	81
	GTO f a	22 31 11
	f LBL 9	31 25 09
	RCL 0	34 00
	h ST I	35 33
	GTO 1	22 01
	f LBL 3	31 25 03
060	h RC I	35 34
	h x≷y	35 52
	÷	81
	RCL 8	34 08
	-	51
	f P≷S	31 42
	h RTN	35 22
	f LBL 1	31 25 01
	RCL 0	34 00
	f P≷S	31 42
070	RCL 9	34 09
	x	71
	STO D	33 14
	RCL 5	34 05
	RCL B	34 12
	÷	81
	3	03
	.	83
	7	07
	x	71
080	STO 6	33 06
	f LOG	31 53
	4	04
	x	71
	STO A	33 11
	RCL 6	34 06
	RCL C	34 13
	h 1/x	35 62
	h y^x	35 63
	1	01
090	.	83
	4	04
	x	71
	RCL A	34 11
	x	71
	STO E	33 15
	RCL D	34 14
	h x≷y	35 52
	g x≤y	32 71
	GTO 4	22 04
100	RCL A	34 11
	GTO f b	22 31 12
	g LBLf e	32 25 15
	.	83
	0	00
	1	01
	f %	31 82
	STO 3	33 03
	2	02
	÷	81
110	-	51
	STO 1	33 01
	STO 0	33 00

STEP	KEY ENTRY	KEY CODE
	f GSB 8	31 22 08
	STO 4	33 04
	RCL 1	34 01
	RCL 3	34 03
	+	61
	STO 0	33 00
	f GSB 8	31 22 08
120	STO 2	33 02
	RCL 4	34 04
	-	51
	RCL 3	34 03
	÷	81
	GTO f d	22 31 14
	f LBL B	32 25 12
	f FIX	31 23
	GTO f e	22 31 15
	f LBL D	32 25 14
130	RCL 2	34 02
	GTO 2	22 02
	f LBL 5	31 25 05
	RCL 0	34 00
	f GSB 8	31 22 08
	STO 2	33 02
	f LBL 2	31 25 02
	RCL 1	34 01
	RCL 0	34 00
	STO 1	33 01
140	-	51
	RCL 4	34 04
	RCL 2	34 02
	STO 4	33 04
	-	51
	÷	81
	x	71
	STO -0	33 51 00
	RCL 0	34 00
	h PAUSE	35 72
150	÷	81
	f RND	31 24
	f x≠0	31 61
	GTO 5	22 05
	RCL 0	34 00
	GTO 3	22 03
	h RTN	35 22
	f LBL 8	31 25 08
	RCL D	34 14
	1	01
160	.	83
	4	04
	÷	81
	RCL 0	34 00
	÷	81
	f LOG	31 53
	RCL C	34 13
	x	71
	4	04

STEP	KEY ENTRY	KEY CODE
	x	71
170	RCL 0	34 00
	-	51
	h RTN	35 22
	f LBL 4	31 25 04
	RCL A	34 11
	GTO 3	22 03
	h RTN	35 22
	f LBL A	31 25 11
	f P≷S	31 42
	RCL 1	34 01
180	EEX	43
	CHS	42
	6	06
	ENTER ↑	41
	h R ↓	35 53
	x	71
	STO B	33 12
	RCL 2	34 02
	STO C	33 13
	.	83
190	8	08
	7	07
	1	01
	RCL 5	34 05
	2	02
	.	83
	6	06
	7	07
	h y^x	35 63
	x	71
200	f P≷S	31 42
	RCL 2	34 02
	g x^2	32 54
	RCL 3	34 03
	g x^2	32 54
	-	51
	RCL 6	34 06
	÷	81
	f$\sqrt{x}$	31 54
	x	71
210	h RTN	35 22
	f LBL C	31 25 13
	RCL D	34 14
	f -x-	31 84
	RCL E	34 15
	f -x-	31 84
	RCL A	34 11
	f -x-	31 84
	f P≷S	31 42
	RCL 0	34 00
220	f -x-	31 84
	f P≷S	31 42
	RCL 0	34 00
	f -x-	31 84
	h RTN	35 22

GP-D-002A - Internal Diameter Computation
&
GP-D-002B

Input:

(1) flowrate - mcfd
(2) constant - 1.0
(3) upstream pressure - psig
(4) downstream pressure - psig
(5) barometric pressure - psia
(6) specific gravity - n
(7) length - miles
(8) flowing temperature - °F
(9) base pressure - psia
(10) base temperature - °F
(11) viscosity - lb.m./ft.sec.
(12) effective roughness - micro inches
(13) bend index - n
(14) elevation difference - feet

Computes:

(1) average line pressure - psig
(2) critical pressure - psia
(3) critical temperature - °R
(4) reduced pressure - n
(5) reduced temperature - n
(6) compressibility - n
(7) α factor
(8) Weymouth Approximation of I.D. - inches
(9) Newton Approximation of I.D. - inches
(10) Reynolds no. at flowing conditions - n
(11) transition Reynolds no. - n

STEP	INSTRUCTIONS	KEYS	OUTPUT UNITS
1.	load card GP-D-002A, side 1 & 2		
2.	input flowrate	A	mcfd
3.	input constant 1.00	A	1.00
4.	input upstream pressure	A	psig
5.	input downstream pressure	A	psig
6.	input barometric pressure	A	psia
7.	input specific gravity	A	n
8.	input length	A	miles
9.	input flowing temperature	A	°F
10.	input base pressure	A	psia
11.	input base temperature	A	°F
12.	input viscosity	A	lb.m./ft.sec.
13.	input effective roughness	A	micro inches
14.	input bend index	A	n
15.	input elevation difference	A	feet
16.	Compute:	B	
	(1) average pressure		psig
	(2) critical pressure		psia

STEP	INSTRUCTIONS	KEYS	OUTPUT UNITS
	(3) critical temperature		^oR
	(4) reduced pressure		n
	(5) reduced temperature		n
	(6) compressibility		n
	(7) α factor		n
17.	load card GP-D-002B, side 1 & 2		
18.	Compute:	A	
	Weymouth Approximation		mcfd
19.	Compute:	E	
	Newton Approximation		mcfd
20.	Reynolds No. at flowing conditions	RCL D	
21.	Transition Reynolds No.	RCL E	
22.	Do you want to run a new case:		
	If yes: (1) f CL REG		
	(2) CL x		0.00

Example: GP-D-002A & B

(1) Calculation of Line Diameter - Fully Turbulent Flow

Given Conditions:

flowrate: 350,000 mcfd
constant: 1.0
inlet pressure: 635 psig
outlet pressure: 485 psig
barometric pressure: 15 psia
specific gravity: .650
line length: 15 miles
flowing temperature: 70^oF
base pressure: 15.025 psia
base temperature: 70^oF
viscosity: 8 x 10^{-6} lb.m./ft.sec.
effective roughness: 950 micro inches
bend index: .96
elevation difference: 0 feet

Execution: Load card GP-D-002A, side 1 & 2

Store parameters in this order:

KEYSTROKES	OUTPUT
350,000 "A"	350,000.000
1.0 "A"	1.000
635 "A"	635.000
485 "A"	485.000
15 "A"	15.000
.650 "A"	.650
15 "A"	15.000

KEYSTROKES	OUTPUT
70 "A"	70.000
15.025 "A"	15.025
70 "A"	70.000
8 x 10^{-6} "A"	8 x 10^{-6}
950 "A"	950.000
.96 "A"	.960
0 "A"	0.000

To Compute:

Press "B"__________"566.15" psig (average pressure)
__________"668.55" psia (critical pressure)
__________"375.00" ^{o}R (critical temperature)
__________".87" (reduced pressure)
__________"1.41" (reduced temperature)
__________".90" (compressibility)
__________"384,949,181.9" (α factor)

Load card GP-D-002B, side 1 & 2

Press "A"__________21.99 inches (Weymouth Approx.)

Press "E"__________21.445 inches (Newton Approx.)

RCL D__________17,950,487.27 (Reynolds No.)

RCL E__________3,691,415.691 (transition Reynolds No.)

Since the Reynolds number is greater than the transition Reynolds number, we have fully turbulent flow.

(2) Calculation of Line Diameter - Partially Turbulent Flow

Given Conditions:

flowrate: 50,000 mcfd
constant: 1.0
inlet pressure: 660 psig
outlet pressure: 650 psig
barometric pressure: 14.9 psia
specific gravity: .625
line length: 17.5 miles
flowing temperature: 70°F
base pressure: 14.9 psia
base temperature: 65°F
viscosity: 8.25 x 10^{-6} lb.m./ft.sec.
effective roughness: 500 micro inches
bend index: .9525
elevation difference: 0 feet

Execution: Clear all registers
Load card GP-D-002A, side 1 & 2

Store parameters in this order:

KEYSTROKES	OUTPUT
50,000 "A"	50,000.0000
1.0 "A"	1.0000
660 "A"	660.0000
650 "A"	650.0000
14.9 "A"	14.9000
.625 "A"	.6250
17.5 "A"	17.5000
70 "A"	70.0000
14.9 "A"	14.9000
65 "A"	65.0000
8.25×10^{-6} "A"	8.25×10^{-6}
500 "A"	500.0000
.9525 "A"	.9525
0 "A"	0.0000

To Compute:

Press "B"	"658.36" psig (average pressure)
	"669.63" psia (critical pressure)
	"367.50" °R (critical temperature)
	"1.01" (reduced pressure)
	"1.44" (reduced temperature)
	".89" (compressibility)
	"51,332,972.58" (α factor)

Load card GP-D-002B

Press "A"	17.607 inches (Weymouth Approx.)
Press "E"	16.933 inches (Newton Approx.)
RCL D	3,031,604.118 (Reynolds No.)
RCL E	6,423,692.491 (transition Reynolds No.)

Since the Reynolds Number at flow conditions is less than the transition Reynolds Number, we have partially turbulent flow.

STEP	KEY ENTRY	KEY CODE
001	f LBL A	31 25 11
	STO (i)	33 24
	f GSB 1	31 22 01
	h RTN	35 22
	f LBL 1	31 25 01
	f ISZ	31 34
	R/S	84
	h ST I	35 33
	GTO 1	22 01
010	h RTN	35 22
	f LBL B	31 25 12
	RCL 2	34 02
	RCL 4	34 04
	+	61
	3	03
	h y^x	35 63
	RCL 3	34 03
	RCL 4	34 04
	+	61
020	3	03
	h y^x	35 63
	−	51
	STO (i)	33 24
	RCL 2	34 02
	RCL 4	34 04
	+	61
	g x^2	32 54
	RCL 3	34 03
	RCL 4	34 04
030	+	61
	g x^2	32 54
	−	51
	RCL (i)	34 24
	h $x \gtrless y$	35 52
	STO (i)	33 24
	÷	81
	.	83
	6	06
	7	07
040	x	71
	RCL 4	34 04
	−	51
	f -x-	31 84
	STO A	33 11
	RCL 5	34 05
	4	04
	3	03
	x	71
	CHS	42
050	6	06
	9	09
	6	06
	.	83
	5	05
	+	61
	STO B	33 12
	f -x-	31 84
	RCL 5	34 05
	3	03
060	0	00
	0	00
	x	71
	1	01
	8	08
	0	00
	+	61
	STO C	33 13
	f -x-	31 84
	RCL A	34 11
070	RCL 4	34 04
	+	61
	RCL B	34 12
	÷	81
	f -x-	31 84
	STO D	33 14
	RCL 7	34 07
	4	04
	6	06
	0	00
080	+	61
	RCL C	34 13
	÷	81
	f -x-	31 84
	STO E	33 15
	1	01
	.	83
	2	02
	5	05
	g $x \leq y$	32 71
090	GTO 2	22 02
	RCL E	34 15
	.	83
	9	09
	x	71
	1	01
	.	83
	3	03
	−	51
	RCL D	34 14
100	x	71
	1	01
	+	61
	f -x-	31 84
	STO B	33 12
	GTO 3	22 03
	f LBL 2	31 25 02
	RCL E	34 15
	.	83
	3	03
110	4	04
	x	71
	.	83

STEP	KEY ENTRY	KEY CODE
	6	06
	-	51
	RCL D	34 14
	x	71
	1	01
	+	61
	STO B	33 12
120	f -x-	31 84
	GTO 3	22 03
	f LBL 3	31 25 03
	RCL (i)	34 24
	f GSB 4	31 22 04
	-	51
	RCL 5	34 05
	÷	81
	RCL 6	34 06
	÷	81
130	RCL 7	34 07
	4	04
	6	06
	0	00
	+	61
	÷	81
	RCL B	34 12
	÷	81
	f$\sqrt{x}$	31 54
	GTO 5	22 05
140	f LBL 4	31 25 04
	RCL A	34 11
	g x^2	32 54
	RCL 5	34 05
	x	71
	f P≷S	31 42
	RCL 3	34 03
	x	71
	f P≷S	31 42
	RCL 7	34 07
150	4	04
	6	06
	0	00
	+	61
	÷	81
	RCL B	34 12
	÷	81
	.	83
	0	00
	3	03
160	7	07
	5	05
	x	71
	h RTN	35 22
	f LBL 5	31 25 05
	3	03
	8	08
	.	83
	7	07
	7	07
170	4	04
	x	71
	RCL 9	34 09
	4	04
	6	06
	0	00
	+	61
	RCL 8	34 08
	÷	81
	x	71
180	1	01
	0	00
	0	00
	0	00
	÷	81
	RCL 0	34 00
	h x≷y	35 52
	÷	81
	f P≷S	31 42
	STO 8	33 08
190	RCL 0	34 00
	f P≷S	31 42
	RCL 9	34 09
	4	04
	6	06
	0	00
	+	61
	x	71
	h 1/x	35 62
	RCL 5	34 05
200	x	71
	RCL 8	34 08
	x	71
	RCL 0	34 00
	x	71
	.	83
	4	04
	7	07
	7	07
	5	05
210	x	71
	f P≷S	31 42
	STO 9	33 09
	f P≷S	31 42
	h RTN	35 22

STEP	KEY ENTRY	KEY CODE
001	f LBL E	31 25 15
	f FIX	31 23
	GTO B	22 12
	g LBLf a	32 25 11
	RCL 2	34 02
	GTO 0	22 00
	f LBL 6	31 25 06
	RCL 0	34 00
	f GSB 9	31 22 09
010	STO 2	33 02
	f LBL 0	31 25 00
	RCL 1	34 01
	RCL 0	34 00
	STO 1	33 01
	-	51
	RCL 4	34 04
	RCL 2	34 02
	STO 4	33 04
	-	51
020	÷	81
	x	71
	STO -0	33 51 00
	RCL 0	34 00
	h PAUSE	35 72
	÷	81
	f RND	31 24
	f x≠0	31 61
	GTO 6	22 06
	RCL 0	34 00
030	h RTN	35 22
	f LBL B	31 25 12
	.	83
	0	00
	1	01
	f %	31 82
	STO 3	33 03
	2	02
	÷	81
	-	51
040	STO 1	33 01
	STO 0	33 00
	f GSB 9	31 22 09
	STO 4	33 04
	RCL 1	34 01
	RCL 3	34 03
	+	61
	STO 0	33 00
	f GSB 9	31 22 09
	STO 2	33 02
050	RCL 4	34 04
	-	51
	RCL 3	34 03
	÷	81
	GTO f a	22 31 11
	f LBL 9	31 25 09
	RCL 0	34 00

STEP	KEY ENTRY	KEY CODE
	f P≷S	31 42
	STO 5	33 05
	f P≷S	31 42
060	h ST I	35 33
	GTO 1	22 01
	f LBL 3	31 25 03
	h RC I	35 34
	2	02
	.	83
	5	05
	h y^x	35 63
	x	71
	RCL 8	34 08
070	-	51
	f P≷S	31 42
	h RTN	35 22
	f LBL 1	31 25 01
	RCL 0	34 00
	f P≷S	31 42
	RCL 9	34 09
	h x≷y	35 52
	÷	81
	STO D	33 14
080	RCL 5	34 05
	RCL B	34 12
	÷	81
	3	03
	.	83
	7	07
	x	71
	STO 6	33 06
	f LOG	31 53
	4	04
090	x	71
	STO A	33 11
	RCL 6	34 06
	RCL C	34 13
	h 1/x	35 62
	h y^x	35 63
	1	01
	.	83
	4	04
	x	71
100	RCL A	34 11
	x	71
	STO E	33 15
	RCL D	34 14
	h x≷y	35 52
	g x≤y	32 71
	GTO 4	22 04
	RCL A	34 11
	GTO f b	22 31 12
	g LBLf e	32 25 15
110	.	83
	0	00
	1	01

STEP	KEY ENTRY	KEY CODE
	f %	31 82
	STO 3	33 03
	2	02
	÷	81
	-	51
	STO 1	33 01
	STO 0	33 00
120	f GSB 8	31 22 08
	STO 4	33 04
	RCL 1	34 01
	RCL 3	34 03
	+	61
	STO 0	33 00
	f GSB 8	31 22 08
	STO 2	33 02
	RCL 4	34 04
	-	51
130	RCL 3	34 03
	÷	81
	GTO f d	22 31 14
	g LBLf b	32 25 12
	f FIX	31 23
	GTO f e	22 31 15
	g LBLf d	32 25 14
	RCL 2	34 02
	GTO 2	22 02
	f LBL 5	31 25 05
140	RCL 0	34 00
	f GSB 8	31 22 08
	STO 2	33 02
	f LBL 2	31 25 02
	RCL 1	34 01
	RCL 0	34 00
	STO 1	33 01
	-	51
	RCL 4	34 04
	RCL 2	34 02
150	STO 4	33 04
	-	51
	÷	81
	x	71
	STO -0	33 51 00
	RCL 0	34 00
	h PAUSE	35 72
	÷	81
	f RND	31 24
	f x≠0	31 61
160	GTO 5	22 05
	RCL 0	34 00
	GTO 3	22 03
	h RTN	35 22
	f LBL 8	31 25 08
	RCL D	34 14
	1	01
	.	83
	4	04
	÷	81
170	RCL 0	34 00
	÷	81
	f LOG	31 53
	RCL C	34 13
	x	71
	4	04
	x	71
	RCL 0	34 00
	-	51
	h RTN	35 22
180	f LBL 4	31 25 04
	RCL A	34 11
	GTO 3	22 03
	h RTN	35 22
	f LBL A	31 25 11
	f P≷S	31 42
	RCL 1	34 01
	EEX	43
	CHS	42
	6	06
190	ENTER ↑	41
	h R↓	35 53
	x	71
	STO B	33 12
	RCL 2	34 02
	STO C	33 13
	f P≷S	31 42
	RCL 6	34 06
	f $\sqrt{x}$	31 54
	RCL 0	34 00
200	x	71
	.	83
	8	08
	7	07
	1	01
	÷	81
	RCL 2	34 02
	g x^2	32 54
	RCL 3	34 03
	g x^2	32 54
210	-	51
	f $\sqrt{x}$	31 54
	÷	81
	.	83
	3	03
	7	07
	5	05
	h y^x	35 63
	h RTN	35 22
	f LBL C	31 25 13
220	RCL D	34 14
	f -x-	31 84
	RCL E	34 15
	f -x-	31 84
	h RTN	35 22

GP-D-003A - Length Computation
&
GP-D-003B

Input:

(1) flowrate - mcfd
(2) viscosity - lb.m./ft.sec.
(3) specific gravity - n
(4) outside diameter - inches
(5) wall thickness - inches
(6) base pressure - psia
(7) base temperature - ^{o}F
(8) effective roughness - micro inches
(9) drag factor - n
(10) elevation difference - feet
(11) inlet pressure - psig
(12) outlet pressure - psig
(13) barometric pressure - psia

Computes:

(1) Reynolds No.
(2) transmission factor, fully turbulent
(3) transmission factor, partially turbulent
(4) critical Reynolds No.
(5) α factor
(6) Weymouth approximation of length
(7) Newton secant approximation of length
(8) critical pressure
(9) critical temperature
(10) reduced temperature
(11) reduced pressure
(12) compressibility

STEP	INSTRUCTIONS	KEYS	OUTPUT UNITS
1.	Load card GP-D-003A, side 1 & 2		
2.	input flowrate	A	mcfd
3.	input viscosity	A	lb.m./ft.sec.
4.	input specific gravity	A	n
5.	input outside diameter	A	inches
6.	input wall thickness	A	inches
7.	input base pressure	A	psia
8.	input base temperature	A	^{o}F
9.	input effective roughness	A	micro inches
10.	input drag factor	A	n
11.	input elevation difference	A	feet
12.	input flowing temperature	A	^{o}F
13.	input inlet pressure	A	psig
14.	input outlet pressure	A	psig
15.	input barometric pressure	A	psia

STEP	INSTRUCTIONS	KEYS	OUTPUT UNITS
16.	Compute:	C	
	(1) Reynolds No.		n
	(2) transmission factor (F_t)		n
	(3) critical Reynolds No.		n
	(4) α factor		n
	(5) transmission factor (P_t)	RCL 0	n (used only in P_t flow)
17.	Load card GP-D-003B, side 1 & 2		
18.	Compute Weymouth approximation	A	miles
19.	Compute Newton secant approximation	B	miles
20.	Compute:	fb	
	(1) critical pressure		psia
	(2) critical temperature		^oR
	(3) reduced temperature		n
	(4) reduced pressure		n
	(5) compressibility		n
21.	Is a new case desired?		
	If yes:		
	(1) f CL REG		
	(2) f P⪌S		
	(3) f CL REG		
	(4) CL x		0.00
22.	Reload card GP-D-003A		

Example: GP-D-003A & B

(1) Calculation of Line Length - Fully Turbulent Flow

Given Conditions:

flowrate: 500,000 mcfd
viscosity: 8.12 x 10^{-6} lb.m./ft.sec.
specific gravity: .610
outside diameter: 30.0 inches
wall thickness: .312 inches
base pressure: 14.696 psia
base temperature: 65°F
effective roughness: 750 micro inches
drag factor: .9700
elevation difference: 0 feet
flowing temperature: 55°F
inlet pressure: 1000 psig
outlet pressure: 725 psig
barometric pressure: 14.80

Execution: Load card GP-D-003A, side 1 & 2

Store parameters in this order:

KEYSTROKES	OUTPUT
500,000 "A"	500,000

KEYSTROKES	OUTPUT
8.12×10^{-6} "A"	8.12×10^{-6}
.610 "A"	.610
30.0 "A"	30.000
.312 "A"	.312
14.696 "A"	14.696
65 "A"	65.000
750 "A"	750.000
.970 "A"	.970
0 "A"	0.000
55 "A"	55.000
1000 "A"	1,000.000
725 "A"	725.000
14.8 "A"	14.800

To Compute:

Press "C" ________ "17,090,877.69" (Reynolds No.)
________ "20.645" (transmission factor) $\sqrt{1/f_{ft}}$
________ "6,049,073.78" (critical Reynolds No.)
________ "4,390.217" (α factor)

Load card GP-D-003B, side 1 & 2

Press "A" ________ 99.38 miles (Weymouth Approx.)

Press "B" ________ 130.34 miles (Newton secant Approx.)

Press "fb" ________ "670.27" psia (critical pressure)
________ "363.00" ^{o}R (critical temperature)
________ "1.42" (reduced temperature)
________ "1.30" (reduced pressure)
________ ".84" (compressibility)

(2) Calculation of Line Length - Partially Turbulent Flow

Given Conditions:

flowrate: 80,000 mcfd
viscosity: 7.8×10^{-6} lb.m./ft.sec.
specific gravity: .6
outside diameter: 24 inches
wall thickness: .250 inches
base pressure: 14.73 psia
base temperature: 60°F
effective roughness: 300 micro inches
drag factor: .9765
elevation difference: 0 feet
flowing temperature: 60°F
inlet pressure: 950 psig
outlet pressure: 925 psig
barometric pressure: 15.00

Execution: Load card GP-D-003A, side 1 & 2

Store parameters in this order:

KEYSTROKES	OUTPUT
80,000 "A"	80,000.0000
7.8 x 10^{-6} "A"	7.8 x 10^{-6}
.6 "A"	.6000
24 "A"	24.0000
.250 "A"	.2500
14.73 "A"	14.7300
60 "A"	60.0000
300 "A"	300.0000
.9765 "A"	.9765
0 "A"	0.0000
60 "A"	60.0000
950 "A"	950.0000
925 "A"	925.0000
15.00 "A"	15.0000

To Compute:

Press "C"	"3,542,024.426" (Reynolds No.)
	"21.849" (transmission factor $_{ft}$)
	"11,999,120.03" (critical Reynolds No.)
	"374.189" (α factor)
RCL 0	19.934 (transmission factor $_{pt}$)
Load card GP-D-003B, side 1 & 2	
Press "A"	116.49 miles (Weymouth Approx.)
Press "B"	150.90 miles (Newton secant Approx.)
Press "fb"	"670.7" psia (critical pressure)
	"360.00" ^{o}R (critical temperature)
	"1.44" (reduced temperature)
	"1.40" (reduced pressure)
	".84" (compressibility)

New case desired? If yes:

f CL REG
f P⪌S
f CL REG
CL x ________ 0.00

STEP	KEY ENTRY	KEY CODE
001	f LBL A	31 25 11
	STO (i)	33 24
	f GSB 1	31 22 01
	h RTN	35 22
	f LBL 1	31 25 01
	f ISZ	31 34
	R/S	84
	h ST I	35 33
	GTO 1	22 01
010	h RTN	35 22
	f LBL B	31 25 12
	2	02
	5	05
	.	83
	7	07
	9	09
	RCL 0	34 00
	RCL 5	34 05
	x	71
020	x	71
	RCL 6	34 06
	4	04
	6	06
	0	00
	+	61
	÷	81
	RCL A	34 11
	2	02
	.	83
030	5	05
	h y^x	35 63
	÷	81
	g x^2	32 54
	RCL 2	34 02
	f P≷S	31 42
	x	71
	RCL 0	34 00
	4	04
	6	06
040	0	00
	+	61
	x	71
	STO 8	33 08
	f P≷S	31 42
	.	83
	0	00
	3	03
	7	07
	5	05
050	RCL 9	34 09
	RCL 2	34 02
	x	71
	x	71
	f P≷S	31 42
	RCL 0	34 00
	4	04
	6	06
	0	00
	+	61
060	÷	81
	STO 9	33 09
	f P≷S	31 42
	h RTN	35 22
	f LBL C	31 25 13
	RCL 0	34 00
	f P≷S	31 42
	STO 4	33 04
	f P≷S	31 42
	RCL 1	34 01
070	÷	81
	RCL 2	34 02
	h ST I	35 33
	x	71
	RCL 3	34 03
	RCL 4	34 04
	2	02
	x	71
	-	51
	STO A	33 11
080	÷	81
	RCL 5	34 05
	x	71
	RCL 6	34 06
	4	04
	6	06
	0	00
	+	61
	÷	81
	.	83
090	4	04
	7	07
	7	07
	5	05
	x	71
	f -x-	31 84
	STO B	33 12
	f GSB B	31 22 12
	RCL A	34 11
	RCL 7	34 07
100	EEX	43
	CHS	42
	6	06
	ENTER ↑	41
	h R↓	35 53
	x	71
	÷	81
	3	03
	.	83
	7	07
110	x	71
	STO D	33 14
	f LOG	31 53

STEP	KEY ENTRY	KEY CODE
	4	04
	x	71
	STO C	33 13
	f -x-	31 84
	RCL D	34 14
	RCL 8	34 08
	h 1/x	35 62
120	h y^x	35 63
	RCL C	34 13
	x	71
	1	01
	.	83
	4	04
	x	71
	STO D	33 14
	f -x-	31 84
	RCL B	34 12
130	h x≷y	35 52
	g x≤y	32 71
	GTO 4	22 04
	RCL C	34 13
	GTO f b	22 31 12
	g LBLf e	32 25 15
	.	83
	0	00
	1	01
	f %	31 82
140	STO 3	33 03
	2	02
	÷	81
	-	51
	STO 1	33 01
	STO 0	33 00
	f GSB 8	31 22 08
	STO 4	33 04
	RCL 1	34 01
	RCL 3	34 03
150	+	61
	STO 0	33 00
	f GSB 8	31 22 08
	STO 2	33 02
	RCL 4	34 04
	-	51
	STO 3	34 03
	÷	81
	GTO f d	22 31 14
	g LBLf b	32 25 12
160	f FIX	31 23
	GTO f e	22 31 15
	g LBLf d	32 25 14
	RCL 2	34 02
	GTO 2	22 02
	f LBL 5	31 25 05
	RCL 0	34 00
	f GSB 8	31 22 08
	STO 2	33 02
	f LBL 2	31 25 02
170	RCL 1	34 01
	RCL 0	34 00
	STO 1	33 01
	-	51
	RCL 4	34 04
	RCL 2	34 02
	STO 4	33 04
	-	51
	÷	81
	x	71
180	STO -0	33 51 00
	RCL 0	34 00
	h PAUSE	35 72
	÷	81
	f RND	31 24
	f x≠0	31 61
	GTO 5	22 05
	RCL 0	34 00
	h 1/x	35 62
	g x^2	32 54
190	f P≷S	31 42
	RCL 8	34 08
	x	71
	STO 8	33 08
	f P≷S	31 42
	h RTN	35 22
	f LBL 8	31 25 08
	RCL B	34 12
	1	01
	.	83
200	4	04
	÷	81
	RCL 0	34 00
	÷	81
	f LOG	31 53
	RCL 8	34 08
	x	71
	4	04
	x	71
	RCL 0	34 00
210	-	51
	h RTN	35 22
	f LBL 4	31 25 04
	RCL C	34 13
	h 1/x	35 62
	g x^2	32 54
	f P≷S	31 42
	RCL 8	34 08
	x	71
	STO 8	33 08
220	f P≷S	31 42
	h RTN	35 22

STEP	KEY ENTRY	KEY CODE
001	f LBL A	31 25 11
	.	83
	8	08
	7	07
	1	01
	f P≷S	31 42
	RCL 4	34 04
	÷	81
	RCL A	34 11
010	2	02
	.	83
	6	06
	7	07
	h y^x	35 63
	x	71
	RCL 1	34 01
	g x^2	32 54
	RCL 2	34 02
	g x^2	32 54
020	-	51
	f$\sqrt{x}$	31 54
	x	71
	g x^2	32 54
	f -x-	31 84
	STO 5	33 05
	f P≷S	31 42
	h RTN	35 22
	f LBL C	31 25 13
	h RC I	35 34
030	4	04
	3	03
	x	71
	CHS	42
	6	06
	9	09
	6	06
	.	83
	5	05
	+	61
040	STO A	33 11
	h RC I	35 34
	3	03
	0	00
	0	00
	x	71
	1	01
	8	08
	0	00
	+	61
050	STO B	33 12
	f P≷S	31 42
	RCL 0	34 00
	f P≷S	31 42
	4	04
	6	06
	0	00
	+	61
	RCL B	34 12
	÷	81
060	STO C	33 13
	h RTN	35 22
	f LBL D	31 25 14
	f P≷S	31 42
	RCL 1	34 01
	RCL 3	34 03
	+	61
	3	03
	h y^x	35 63
	RCL 2	34 02
070	RCL 3	34 03
	+	61
	3	03
	h y^x	35 63
	-	51
	STO 7	33 07
	RCL 1	34 01
	RCL 3	34 03
	+	61
	g x^2	32 54
080	RCL 2	34 02
	RCL 3	34 03
	+	61
	g x^2	32 54
	-	51
	RCL 7	34 07
	h x≷y	35 52
	÷	81
	.	83
	6	06
090	7	07
	x	71
	RCL 3	34 03
	-	51
	f P≷S	31 42
	STO D	33 14
	h RTN	35 22
	f LBL E	31 25 15
	RCL D	34 14
	RCL A	34 11
100	h PAUSE	35 72
	÷	81
	STO E	33 15
	RCL C	34 13
	.	83
	3	03
	4	04
	1	01
	x	71
	.	83
110	6	06
	0	00
	4	04

STEP	KEY ENTRY	KEY CODE
	CHS	42
	+	61
	RCL E	34 15
	x	71
	1	01
	+	61
	STO E	33 15
120	h RTN	35 22
	f LBL B	31 25 12
	f GSB C	31 22 13
	f GSB D	31 22 14
	f GSB E	31 22 15
	f P≷S	31 42
	RCL 1	34 01
	RCL 3	34 03
	+	61
	g x^2	32 54
130	RCL 2	34 02
	RCL 3	34 03
	+	61
	g x^2	32 54
	-	51
	STO 7	33 07
	RCL D	34 14
	g x^2	32 54
	RCL E	34 15
	÷	81
140	RCL 9	34 09
	x	71
	CHS	42
	RCL 7	34 07
	+	61
	RCL 8	34 08
	÷	81
	RCL E	34 15
	÷	81
	f P≷S	31 42
150	f -x-	31 84
	h RTN	35 22
	g LBLf b	32 25 12
	RCL A	34 11
	f -x-	31 84
	RCL B	34 12
	f -x-	31 84
	RCL C	34 13
	f -x-	31 84
	RCL D	34 14
160	RCL A	34 11
	÷	81
	f -x-	31 84
	RCL E	34 15
	h RTN	35 22

GP-D-004A - Upstream Pressure Computation
&
GP-D-004B

Input:

(1) flowrate - mcfd
(2) viscosity - lb.m./ft.sec.
(3) specific gravity - n
(4) outside diameter - inches
(5) wall thickness - inches
(6) base pressure - psia
(7) base temperature - ^{o}F
(8) effective roughness - micro inches
(9) drag factor - n
(10) elevation difference - feet
(11) line length - miles
(12) flowing temperature - ^{o}F
(13) downstream pressure - psia
(14) barometric pressure - psia

Computes:

(1) Reynolds No.
(2) transmission factor - fully turbulent
(3) transmission factor - partially turbulent
(4) critical Reynolds No.
(5) α factor
(6) Weymouth approximation of upstream pressure
(7) Newton approximation of upstream pressure
(8) critical pressure
(9) critical temperature
(10) reduced temperature
(11) average pressure
(12) compressibility

STEP	INSTRUCTIONS	KEYS	OUTPUT UNITS
1.	load card GP-D-004A, side 1 & 2		
2.	input flowrate	A	mcfd
3.	input viscosity	A	lb.m./ft.sec.
4.	input specific gravity	A	n
5.	input outside diameter	A	inches
6.	input wall thickness	A	inches
7.	input base pressure	A	psia
8.	input base temperature	A	^{o}F
9.	input effective roughness	A	micro inches
10.	input drag factor	A	n
11.	input elevation difference	A	feet
12.	input line length	A	miles
13.	input flowing temperature	A	^{o}F
14.	input downstream pressure	A	psig
15.	input barometric pressure	A	psia

STEP	INSTRUCTIONS	KEYS	OUTPUT UNITS
16.	Compute:	C	
	(1) Reynolds No.		n
	(2) transmission factor $\sqrt{1/f_{ft}}$		n
	(3) critical Reynolds No.		n
	(4) α factor		n
	(5) transmission factor $\sqrt{1/f_{pt}}$	RCL 0	n (only for P_t flow)
17.	load card GP-D-004B, side 1 & 2		
18.	Compute Upstream Pressure:	B	
	(1) Weymouth Approximation		psig
	(2) Newton secant Approximation		psig
19.	Do you wish to run another case		
	If yes:		
	f CL REG		
	f P≷S		
	f CL REG		
	CL x		0.00

Example: GP-D-004 A & B

(1) Upstream Pressure Computation - Fully Turbulent Flow

Given Conditions:

flowrate: 350,000 mcfd
viscosity: 7.97×10^{-6} lb.m./ft.sec.
specific gravity: .595
outside diameter: 24 inches
wall thickness: .250 inches
base pressure: 14.4 psia
base temperature: 65°F
effective roughness: 500 micro inches
drag factor: .965
elevation difference: 0 feet
line length: 12 miles
flowing temperature: 60°F
downstream pressure: 400 psig
barometric pressure: 14.5 psia

Execution: Load card GP-D-004A, side 1 & 2

Store parameters in this order:

KEYSTROKES	OUTPUT
350,000 "A"	350,000.000
7.97×10^{-6} "A"	7.97×10^{-6}
.595 "A"	.595
24 "A"	24.000
.250 "A"	.250
14.4 "A"	14.400
65 "A"	65.000

KEYSTROKES	OUTPUT
500 "A"	500.000
.965 "A"	.965
0 "A"	0.000
12 "A"	12.000
60 "A"	60.000
400 "A"	400.000
14.5 "A"	14.500

To Compute:

Press "C"__________"14,562,481.65" (Reynolds No.)
__________"20.961" ($\sqrt{1/f_{ft}}$)
__________"7,905,082.445" (critical Reynolds No.)
__________"72,272,.85" (α factor)

Load card GP-D-004B, side 1 & 2

Press "B" to compute upstream pressure
__________"502.42" psig (Weymouth Approximation)
__________"474.27" psig (Newton Approximation)

RCL A__________670.92 psia (critical pressure)

RCL B__________358.50^{o}R (critical temperature)

RCL C__________1.45 (reduced temperature)

RCL D__________440.42 psig (average pressure)

RCL E__________.93 (compressibility)

(2) Upstream Pressure Computation - Partially Turbulent Flow

Given Conditions:

flowrate: 60,000 mcfd
viscosity: 8 x 10^{-6} lb.m./ft.sec.
specific gravity: .582
outside diameter: 12 inches
wall thickness: .250 inches
base pressure: 14.7 psia
base temperature: 62°F
effective roughness: 300 micro inches
drag factor: .966
elevation difference: 0 feet
line length: 20 miles
flowing temperature: 69°F
downstream pressure: 932.86 psig
barometric pressure: 14.7 psia

Execution: Load card GP-D-004A, side 1 & 2

Store parameters in this order:

KEYSTROKES	OUTPUT
60,000 "A"	60,000.000
8×10^{-6} "A"	8×10^{-6}
.582 "A"	.582
12 "A"	12.000
.250 "A"	.250
14.7 "A"	14.700
62 "A"	62.000
300 "A"	300.000
.966 "A"	.966
0 "A"	0.000
20 "A"	20.000
69 "A"	69.000
932.86 "A"	932.860
14.7 "A"	14.700

To Compute:

Press "C"	"5,103,952.4" (Reynolds No.)
	"20.607" ($\sqrt{1/f_{ft}}$)
	"6,212,251.529" (critical Reynolds No.)
	"141,034.556" (α factor)
RCL 0	20.302 ($\sqrt{1/f_{pt}}$)

Load card CP-D-004B, side 1 & 2

Press "B" to compute upstream pressure

	"1037.25" psig (Weymouth Approximation)
	"995.01"psig(Newton Approximation)
RCL A	671.47 psia (critical pressure)
RCL B	354.60 ^{o}R (critical temperature)
RCL C	1.49 (reduced temperature)
RCL D	969.16 psig (average pressure)
RCL E	.86 (compressibility)

STEP	KEY ENTRY	KEY CODE
001	f LBL A	31 25 11
	STO (i)	33 24
	f GSB 1	31 22 01
	h RTN	35 22
	f LBL 1	31 25 01
	f ISZ	31 34
	R/S	84
	h ST I	35 33
	GTO 1	22 01
010	h RTN	35 22
	f LBL B	31 25 12
	2	02
	5	05
	.	83
	7	07
	9	09
	RCL 0	34 00
	RCL 5	34 05
	x	71
020	x	71
	RCL 6	34 06
	4	04
	6	06
	0	00
	+	61
	÷	81
	RCL A	34 11
	2	02
	.	83
030	5	05
	h y^x	35 63
	÷	81
	g x^2	32 54
	RCL 2	34 02
	f P≷S	31 42
	RCL 0	34 00
	x	71
	x	71
	RCL 1	34 01
040	4	04
	6	06
	0	00
	+	61
	x	71
	STO 8	33 08
	f P≷S	31 42
	.	83
	0	00
	3	03
050	7	07
	5	05
	RCL 9	34 09
	RCL 2	34 02
	x	71
	x	71
	f P≷S	31 42

STEP	KEY ENTRY	KEY CODE
	RCL 1	34 01
	4	04
	6	06
060	0	00
	+	61
	÷	81
	STO 9	33 09
	f P≷S	31 42
	h RTN	35 22
	f LBL C	31 25 13
	RCL 0	34 00
	f P≷S	31 42
	STO 4	33 04
070	f P≷S	31 42
	RCL 1	34 01
	÷	81
	RCL 2	34 02
	h ST I	35 33
	x	71
	RCL 3	34 03
	RCL 4	34 04
	2	02
	x	71
080	-	51
	STO A	33 11
	÷	81
	RCL 5	34 05
	x	71
	RCL 6	34 06
	4	04
	6	06
	0	00
	+	61
090	÷	81
	.	83
	4	04
	7	07
	7	07
	5	05
	x	71
	f -x-	31 84
	STO B	33 12
	f GSB B	31 22 12
100	RCL A	34 11
	RCL 7	34 07
	EEX	43
	CHS	42
	6	06
	ENTER ↑	41
	h R ↓	35 53
	x	71
	÷	81
	3	03
110	.	83
	7	07
	x	71

STEP	KEY ENTRY	KEY CODE
	STO D	33 14
	f LOG	31 53
	4	04
	x	71
	STO C	33 13
	f -x-	31 84
	RCL D	34 14
120	RCL 8	34 08
	h 1/x	35 62
	h y^x	35 63
	RCL C	34 13
	x	71
	1	01
	.	83
	4	04
	x	71
	STO D	33 14
130	f -x-	31 84
	RCL B	34 12
	h x≷y	35 52
	g x≤y	32 71
	GTO 4	22 04
	RCL C	34 13
	GTO f b	22 31 12
	g LBLf e	32 25 15
	.	83
	0	00
140	1	01
	f %	31 82
	STO C	33 03
	2	02
	÷	81
	-	51
	STO 1	33 01
	STO 0	33 00
	f GSB 8	31 22 08
	STO 4	33 04
150	RCL 1	34 01
	RCL 3	34 03
	+	61
	STO 0	33 00
	f GSB 8	31 22 08
	STO 2	33 02
	RCL 4	34 04
	-	51
	RCL 3	34 03
	÷	81
160	GTO f d	22 31 14
	g LBLf b	32 25 12
	f FIX	31 23
	GTO f e	22 31 15
	g LBLf d	32 25 14
	RCL 2	34 02
	GTO 2	22 02
	f LBL 5	31 25 05
	RCL 0	34 00

STEP	KEY ENTRY	KEY CODE
	f GSB 8	31 22 08
170	STO 2	33 02
	f LBL 2	31 25 02
	RCL 1	34 01
	RCL 0	34 00
	STO 1	33 01
	-	51
	RCL 4	34 04
	RCL 2	34 02
	STO 4	33 04
	-	51
180	÷	81
	x	71
	STO -0	33 51 00
	RCL 0	34 00
	h PAUSE	35 72
	÷	81
	f RND	31 24
	f x≠0	31 61
	GTO 5	22 05
	RCL 0	34 00
190	h 1/x	35 62
	g x^2	32 54
	f P≷S	31 42
	RCL 8	34 08
	x	71
	STO 8	33 08
	f P≷S	31 42
	h RTN	35 22
	f LBL 8	31 25 08
	RCL B	34 12
200	1	01
	.	83
	4	04
	÷	81
	RCL 0	34 00
	÷	81
	f LOG	31 53
	RCL 8	34 08
	x	71
	4	04
210	x	71
	RCL 0	34 00
	-	51
	h RTN	35 22
	f LBL 4	31 25 04
	RCL C	34 13
	h 1/x	35 62
	g x^2	32 54
	f P≷S	31 42
	RCL 8	34 08
220	x	71
	STO 8	33 08
	f P≷S	31 42
	h RTN	35 22

STEP	KEY ENTRY	KEY CODE
001	f LBL A	31 25 11
	f P≷S	31 42
	RCL 4	34 04
	RCL 0	34 00
	f $\sqrt{x}$	31 54
	x	71
	.	83
	8	08
	7	07
010	1	01
	÷	81
	RCL 5	34 05
	2	02
	.	83
	6	06
	7	07
	h y^x	35 63
	÷	81
	g x^2	32 54
020	RCL 2	34 02
	g x^2	32 54
	+	61
	f $\sqrt{x}$	31 54
	f -x-	31 84
	STO 5	33 05
	f P≷S	31 42
	h RTN	35 22
	f LBL C	31 25 13
	h RC I	35 34
030	4	04
	3	03
	x	71
	CHS	42
	6	06
	9	09
	6	06
	.	83
	5	05
	+	61
040	STO A	33 11
	h RC I	35 34
	3	03
	0	00
	0	00
	x	71
	1	01
	8	08
	0	00
	+	61
050	STO B	33 12
	f P≷S	31 42
	RCL 1	34 01
	f P≷S	31 42
	4	04
	6	06
	0	00

STEP	KEY ENTRY	KEY CODE
	+	61
	RCL B	34 12
	÷	81
060	STO C	33 13
	h RTN	35 22
	f LBL D	31 25 14
	f P≷S	31 42
	RCL 2	34 02
	RCL 3	34 03
	+	61
	3	03
	h y^x	35 63
	f P≷S	31 42
070	RCL 0	34 00
	f P≷S	31 42
	RCL 3	34 03
	+	61
	3	03
	h y^x	35 63
	h x≷y	35 52
	-	51
	STO 7	33 07
	RCL 2	34 02
080	RCL 3	34 03
	+	61
	g x^2	32 54
	RCL 3	34 03
	f P≷S	31 42
	RCL 0	34 00
	+	61
	g x^2	32 54
	h x≷y	35 52
	-	51
090	f P≷S	31 42
	RCL 7	34 07
	h x≷y	35 52
	÷	81
	.	83
	6	06
	7	07
	x	71
	RCL 3	34 03
	-	51
100	f P≷S	31 42
	STO D	33 14
	h RTN	35 22
	f LBL E	31 25 15
	RCL D	34 14
	RCL A	34 11
	h PAUSE	35 72
	÷	81
	STO E	33 15
	RCL C	34 13
110	.	83
	3	03
	4	04

STEP	KEY ENTRY	KEY CODE
	1	01
	x	71
	.	83
	6	06
	0	00
	4	04
	CHS	42
120	+	61
	RCL E	34 15
	x	71
	1	01
	+	61
	STO E	33 15
	h RTN	35 22
	g LBLf b	32 25 12
	f P$\gtrless$S	31 42
	RCL 8	34 08
130	RCL E	34 15
	x	71
	RCL 2	34 02
	RCL 3	34 03
	+	61
	g x^2	32 54
	+	61
	STO 6	33 06
	RCL 9	34 09
	RCL D	34 14
140	g x^2	32 54
	x	71
	RCL E	34 15
	÷	81
	RCL 6	34 06
	+	61
	f$\sqrt{x}$	31 54
	CHS	42
	f P$\gtrless$S	31 42
	RCL 0	34 00
150	f P$\gtrless$S	31 42
	RCL 3	34 03
	+	61
	+	61
	f P$\gtrless$S	31 42
	h RTN	35 22
	f LBL B	31 25 12
	RCL A	34 11
	f P$\gtrless$S	31 42
	STO 5	33 05
160	f P$\gtrless$S	31 42
	f GSB C	31 22 13
	f GSB A	31 22 11
	GTO f b	22 31 12
	g LBLf e	32 25 15
	.	83
	0	00
	1	01
	f %	31 82
	STO 3	33 03
170	2	02
	÷	81
	−	51
	STO 1	33 01
	STO 0	33 00
	f GSB 8	31 22 08
	STO 4	33 04
	RCL 1	34 01
	RCL 3	34 03
	+	61
180	STO 0	33 00
	f GSB 8	31 22 08
	STO 2	33 02
	RCL 4	34 04
	−	51
	RCL 3	34 03
	÷	81
	GTO f d	22 31 14
	g LBLf b	32 25 12
	f FIX	31 23
190	GTO f e	22 31 15
	g LBLf d	32 25 14
	RCL 2	34 02
	GTO 2	22 02
	f LBL 5	31 25 05
	RCL 0	34 00
	f GSB 8	31 22 08
	STO 2	33 02
	f LBL 2	31 25 02
	RCL 1	34 01
200	RCL 0	34 00
	STO 1	33 01
	−	51
	RCL 4	34 04
	RCL 2	34 02
	STO 4	33 04
	−	51
	÷	81
	x	71
	STO −0	33 51 00
210	RCL 0	34 00
	h PAUSE	35 72
	÷	81
	f RND	31 24
	f x≠0	31 61
	GTO 5	22 05
	RCL 0	34 00
	h RTN	35 22
	f LBL 8	31 25 08
	f GSB D	31 22 14
220	f GSB E	31 22 15
	g GSB b	32 22 12
	h RTN	35 22

GP-D-005A - Downstream Pressure Computation
&
GP-D-005B

Input:

(1) flowrate - mcfd
(2) viscosity - lb.m./ft.sec.
(3) specific gravity - n
(4) outside diameter - inches
(5) wall thickness - inches
(6) base pressure - psia
(7) base temperature - ^oF
(8) effective roughness - micro inches
(9) drag factor - n
(10) elevation difference - feet
(11) line length - miles
(12) flowing temperature - ^oF
(13) upstream pressure - psig
(14) barometric pressure - psia

Computes:

(1) Reynolds No.
(2) transmission factor - fully turbulent
(3) transmission factor - partially turbulent
(4) critical Reynolds No.
(5) α factor
(6) Weymouth approximation of downstream pressure
(7) Newton approximation of downstream pressure
(8) critical pressure
(9) critical temperature
(10) reduced temperature
(11) average pressure
(12) compressibility

STEP	INSTRUCTIONS	KEYS	OUTPUT UNITS
1.	load card GP-D-005A, side 1 & 2		
2.	input flowrate	A	mcfd
3.	input viscosity	A	lb.m./ft.sec
4.	input specific gravity	A	n
5.	input outside diameter	A	inches
6.	input wall thickness	A	inches
7.	input base pressure	A	psia
8.	input base temperature	A	^oF
9.	input effective roughness	A	micro inches
10.	input drag factor	A	n
11.	input elevation difference	A	feet
12.	input line length	A	miles
13.	input flowing temperature	A	^oF
14.	input upstream pressure	A	psig
15.	input barometric pressure	A	psia

STEP	INSTRUCTIONS	KEYS	OUTPUT UNITS
16.	Compute:	C	
	(1) Reynolds No.		n
	(2) transmission factor $\sqrt{1/f_{ft}}$		n
	(3) critical Reynolds No.		n
	(4) α factor		n
	(5) transmission factor $\sqrt{1/f_{pt}}$	RCL 0	n (only in P_t flow)
17.	load card GP-D-005B, side 1 & 2		
18.	Compute Downstream Pressure:	B	
	(1) Weymouth Approximation		psig
	(2) Newton secant Approximation		psig
19.	Do you wish to run another case		
	If yes:		
	f CL REG		
	f P$\gtrless$S		
	f CL REG		
	CL x		0.00

Example: GP-D-005 A & B

(1) Downstream Pressure Computation - Fully Turbulent Flow

Given Conditions:

flowrate: 350,000 mcfd
viscosity: 7.97 x 10^{-6} lb.m./ft.sec.
specific gravity: .595
outside diameter: 24 inches
wall thickness: .250 inches
base pressure: 14.4 psia
base temperature: 65 ^{o}F
effective roughness: 500 micro inches
drag factor: .965
elevation difference: 0 feet
line length: 12 miles
flowing temperature: 60°F
upstream pressure: 467.10 psig
barometric pressure: 14.5 psia

Execution: Load card GP-D-005A, side 1 & 2

Store parameters in this order:

KEYSTROKES	OUTPUT
350,000 "A"	350,000.000
7.97 x 10^{-6} "A"	7.97 x 10^{-6}
.595 "A"	.595
24 "A"	24.000
.250 "A"	.250
14.4 "A"	14.400
65 "A"	65.000

KEYSTROKES	OUTPUT
500 "A"	500.000
.965 "A"	.965
0 "A"	0.000
12.0 "A"	12.000
60 "A"	60.000
474.27 "A"	474.270
14.5 "A"	14.500

To Compute:

Press "C"	"14,562,481.65" (Reynolds No.)
	"20.961" ($\sqrt{1/f_{ft}}$)
	"7,905,082.45" (critical Reynolds No.)
	"72,272.85" (α factor)

Load card GP-D-005B, side 1 & 2

Press "B" to compute downstream pressure

	"364.02" psig (Weymouth Approximation)
	"400.00" psig (Newton Approximation)
RCL A	670.92 psia (critical pressure)
RCL B	358.50 ^{o}R (critical temperature)
RCL C	1.45 (reduced temperature)
RCL D	440.67 psig (average pressure)
RCL E	.93 (compressibility)

(2) Downstream Pressure Computation - Partially Turbulent Flow

Given Conditions:

flowrate: 60,000 mcfd
viscosity: 8 x 10^{-6} lb.m./ft.sec.
specific gravity: .582
outside diameter: 12 inches
wall thickness: .250 inches
base pressure: 14.7 psia
base temperature: 62°F
effective roughness: 300 micro inches
drag factor: .966
elevation difference: 0 feet
line length: 20 miles
flowing temperature: 69°F
upstream pressure: 995.01 psig
barometric pressure: 14.7 psia

Execution: Load card GP-D-005A, side 1 & 2

Store parameters in this order:

KEYSTROKES	OUTPUT
60,000 "A"	60,000.000
8 x 10^{-6} "A"	8 x 10^{-6}
.582 "A"	.582
12 "A"	12.000
.250 "A"	.250
14.7 "A"	14.700
62 "A"	62.000
300 "A"	300.000
.966 "A"	.966
0 "A"	0.000
20 "A"	20.000
69 "A"	69.000
995.01 "A"	995.010
14.7	14.700

To Compute:

Press "C"	"5,103,952.4" (Reynolds No.)
	"20.607" ($\sqrt{1/f_{ft}}$)
	"6,212,251.529" (critical Reynolds No.)
	"141,034.556 (α factor)
RCL 0	20.302 ($\sqrt{1/f_{pt}}$)

Load card GP-D-005B, side 1 & 2

Press "B" to compute downstream pressure

	"885.65" psig (Weymouth Approximation)
	"932.86" psig (Newton Approximation)
RCL A	671.47 psia (critical pressure)
RCL B	354.60 ^{o}R (critical temperature)
RCL C	1.49 (reduced temperature)
RCL D	969.24 psig (average pressure)
RCL E	.86 (compressibility)

STEP	KEY ENTRY	KEY CODE
001	f LBL A	31 25 11
	STO (i)	33 24
	f GSB 1	31 22 01
	h RTN	35 22
	f LBL 1	31 25 01
	f ISZ	31 34
	R/S	84
	h ST I	35 33
	GTO 1	22 01
010	h RTN	35 22
	f LBL B	31 25 12
	2	02
	5	05
	.	83
	7	07
	9	09
	RCL 0	34 00
	RCL 5	34 05
	x	71
020	x	71
	RCL 6	34 06
	4	04
	6	06
	0	00
	+	61
	÷	81
	RCL A	34 11
	2	02
	.	83
030	5	05
	h y^x	35 63
	÷	81
	g x^2	32 54
	RCL 2	34 02
	f P≷S	31 42
	RCL 0	34 00
	x	71
	x	71
	RCL 1	34 01
040	4	04
	6	06
	0	00
	+	61
	x	71
	STO 8	33 08
	f P≷S	31 42
	.	83
	0	00
	3	03
050	7	07
	5	05
	RCL 9	34 09
	RCL 2	34 02
	x	71
	x	71
	f P≷S	31 42
	RCL 1	34 01
	4	04
	6	06
060	0	00
	+	61
	÷	81
	STO 9	33 09
	f P≷S	31 42
	h RTN	35 22
	f LBL C	31 25 13
	RCL 0	34 00
	f P≷S	31 42
	STO 4	33 04
070	f P≷S	31 42
	RCL 1	34 01
	÷	81
	RCL 2	34 02
	h ST I	35 33
	x	71
	RCL 3	34 03
	RCL 4	34 04
	2	02
	x	71
080	-	51
	STO A	33 11
	÷	81
	RCL 5	34 05
	x	71
	RCL 6	34 06
	4	04
	6	06
	0	00
	+	61
090	÷	81
	.	83
	4	04
	7	07
	7	07
	5	05
	x	71
	f -x-	31 84
	STO B	33 12
	f GSB B	31 22 12
100	RCL A	34 11
	RCL 7	34 07
	EEX	43
	CHS	42
	6	06
	ENTER ↑	41
	h R ↓	35 53
	x	71
	÷	81
	3	03
110	.	83
	7	07
	x	71

STEP	KEY ENTRY	KEY CODE
	STO D	33 14
	f LOG	31 53
	4	04
	x	71
	STO C	33 13
	f -x-	31 84
	RCL D	34 14
120	RCL 8	34 08
	h 1/x	35 62
	h y^x	35 63
	RCL C	34 13
	x	71
	1	01
	.	83
	4	04
	x	71
	STO D	33 14
130	f -x-	31 84
	RCL B	34 12
	h x≷y	35 52
	g x≤y	32 71
	GTO 4	22 04
	RCL C	34 13
	GTO f b	22 31 12
	g LBLf e	32 25 15
	.	83
	0	00
140	1	01
	f %	31 82
	STO C	33 03
	2	02
	÷	81
	-	51
	STO 1	33 01
	STO 0	33 00
	f GSB 8	31 22 08
	STO 4	33 04
150	RCL 1	34 01
	RCL 3	34 03
	+	61
	STO 0	33 00
	f GSB 8	31 22 08
	STO 2	33 02
	RCL 4	34 04
	-	51
	RCL 3	34 03
	÷	81
160	GTO f d	22 31 14
	g LBLf b	32 25 12
	f FIX	31 23
	GTO f e	22 31 15
	g LBLf d	32 25 14
	RCL 2	34 02
	GTO 2	22 02
	f LBL 5	31 25 05
	RCL 0	34 00

STEP	KEY ENTRY	KEY CODE
	f GSB 8	31 22 08
170	STO 2	33 02
	f LBL 2	31 25 02
	RCL 1	34 01
	RCL 0	34 00
	STO 1	33 01
	-	51
	RCL 4	34 04
	RCL 2	34 02
	STO 4	33 04
	-	51
180	÷	81
	x	71
	STO -0	33 51 00
	RCL 0	34 00
	h PAUSE	35 72
	÷	81
	f RND	31 24
	f x≠0	31 61
	GTO 5	22 05
	RCL 0	34 00
190	h 1/x	35 62
	g x^2	32 54
	f P≷S	31 42
	RCL 8	34 08
	x	71
	STO 8	33 08
	f P≷S	31 42
	h RTN	35 22
	f LBL 8	31 25 08
	RCL B	34 12
200	1	01
	.	83
	4	04
	÷	81
	RCL 0	34 00
	÷	81
	f LOG	31 53
	RCL 8	34 08
	x	71
	4	04
210	x	71
	RCL 0	34 00
	-	51
	h RTN	35 22
	f LBL 4	31 25 04
	RCL C	34 13
	h 1/x	35 62
	g x^2	32 54
	f P≷S	31 42
	RCL 8	34 08
220	x	71
	STO 8	33 08
	f P≷S	31 42
	h RTN	35 22

STEP	KEY ENTRY	KEY CODE
001	f LBL A	31 25 11
	f P≷S	31 42
	RCL 4	34 04
	RCL 0	34 00
	f$\sqrt{x}$	31 54
	x	71
	.	83
	8	08
	7	07
010	1	01
	÷	81
	RCL 5	34 05
	2	02
	.	83
	6	06
	7	07
	h y^x	35 63
	÷	81
	g x^2	32 54
020	CHS	42
	RCL 2	34 02
	g x^2	32 54
	+	61
	f$\sqrt{x}$	31 54
	f -x-	31 84
	STO 5	33 05
	f P≷S	31 42
	h RTN	35 22
	f LBL C	31 25 13
030	h RC I	35 34
	4	04
	3	03
	x	71
	CHS	42
	6	06
	9	09
	6	06
	.	83
	5	05
040	+	61
	STO A	33 11
	h RC I	35 34
	3	03
	0	00
	0	00
	x	71
	1	01
	8	08
	0	00
050	+	61
	STO B	33 12
	f P≷S	31 42
	RCL 1	34 01
	f P≷S	31 42
	4	04
	6	06
	0	00
	+	61
	RCL B	34 12
060	÷	81
	STO C	33 13
	h RTN	35 22
	f LBL D	31 25 14
	f P≷S	31 42
	RCL 2	34 02
	RCL 3	34 03
	+	61
	3	03
	h y^x	35 63
070	f P≷S	31 42
	RCL 0	34 00
	f P≷S	31 42
	RCL 3	34 03
	+	61
	3	03
	h y^x	35 63
	-	51
	STO 7	33 07
	RCL 2	34 02
080	RCL 3	34 03
	+	61
	g x^2	32 54
	RCL 3	34 03
	f P≷S	31 42
	RCL 0	34 00
	+	61
	g x^2	32 54
	-	51
	f P≷S	31 42
090	RCL 7	34 07
	h x≷y	35 52
	÷	81
	.	83
	6	06
	7	07
	x	71
	RCL 3	34 03
	-	51
	f P≷S	31 42
100	STO D	33 14
	h RTN	35 22
	f LBL E	31 25 15
	RCL D	34 14
	RCL A	34 11
	h PAUSE	35 72
	÷	81
	STO E	33 15
	RCL C	34 13
	.	83
110	3	03
	4	04
	1	01

STEP	KEY ENTRY	KEY CODE
	x	71
	.	83
	6	06
	0	00
	4	04
	CHS	42
	+	61
120	RCL E	34 15
	x	71
	1	01
	+	61
	STO E	33 15
	h RTN	35 22
	g LBLf b	32 25 12
	f P≷S	31 42
	RCL 8	34 08
	RCL E	34 15
130	x	71
	f P≷S	31 42
	RCL 0	34 00
	f P≷S	31 42
	RCL 3	34 03
	+	61
	g x^2	32 54
	+	61
	STO 6	33 06
	RCL 9	34 09
140	RCL D	34 14
	g x^2	32 54
	x	71
	RCL E	34 15
	÷	81
	RCL 6	34 06
	+	61
	f $\sqrt{x}$	31 54
	CHS	42
	RCL 2	34 02
150	RCL 3	34 03
	+	61
	+	61
	f P≷S	31 42
	h RTN	35 22
	f LBL B	31 25 12
	RCL A	34 11
	f P≷S	31 42
	STO 5	33 05
	f P≷S	31 42
160	f GSB C	31 22 13
	f GSB A	31 22 11
	GTO f b	22 31 12
	g LBLf e	32 25 15
	.	83
	0	00
	1	01
	f %	31 82
	STO 3	33 03
	2	02
170	÷	81
	-	51
	STO 1	33 01
	STO 0	33 00
	f GSB 8	31 22 08
	STO 4	33 04
	RCL 1	34 01
	RCL 3	34 03
	+	61
	STO 0	33 00
180	f GSB 8	31 22 08
	STO 2	33 02
	RCL 4	34 04
	-	51
	RCL 3	34 03
	÷	81
	GTO f d	22 31 14
	g LBLf b	32 25 12
	f FIX	31 23
	GTO f e	22 31 15
190	g LBLf d	32 25 14
	RCL 2	34 02
	GTO 2	22 02
	f LBL 5	31 25 05
	RCL 0	34 00
	f GSB 8	31 22 08
	STO 2	33 02
	f LBL 2	31 25 02
	RCL 1	34 01
	RCL 0	34 00
200	STO 1	33 01
	-	51
	RCL 4	34 04
	RCL 2	34 02
	STO 4	33 04
	-	51
	÷	81
	x	71
	STO -0	33 51 00
	RCL 0	34 00
210	h PAUSE	35 72
	÷	81
	f RND	31 24
	f x≠0	31 61
	GTO 5	22 05
	RCL 0	34 00
	h RTN	35 22
	f LBL 8	31 25 08
	f GSB D	31 22 14
	f GSB E	31 22 15
220	g GSB b	32 22 12
	h RTN	35 22

Nomenclature

BI - bend index - degrees/mile
e - base of natural logarithms, 2.718
F_f - drag factor
F_t - smooth-pipe law transmission factor
f - fanning friction factor
G - gas gravity (air = 1)
*H - elevation - feet
I.D. - internal pipe diameter - inches
k_e - effective or operating roughness - microinches
L - line length - miles
P_b - base pressure - psia
P_m - average pressure - psia
P_1 - upstream pressure - psia
P_2 - downstream pressure - psia
Q_b - volumetric flow rate at base conditions - MCF/day
T - flowing gas temperature - °R(°F + 460)
T_b - base temperature - °R(°F + 460)
z - compressibility factor
μ - dynamic viscosity - lb./ft.-sec.

Subscripts

ft - fully turbulent
pt - partially turbulent
rpl - rough-pipe law
spl - smooth-pipe law

*Note: (1) elevation is positive for net uphill flow.
(2) elevation is negative for net downhill flow.

3.0 Two Phase Flow

3.1 Introduction

There are twelve basic programs for the detailed design hydraulics of two phase flow. They are divided into these principal parts. The Baker method is utilized for these programs.

TPF-D-001 TPF-D-002	Oil Characteristics
TPF-D-003	Gas Characteristics
TPF-D-004 TPF-D-005	I.D. for Oil Flowrate only
TPF-D-006A TPF-D-006B	I.D. for Gas Flowrate only
TPF-D-007	Pressure drop for Oil Flowrate in selected I.D. only
TPF-D-008A TPF-D-008B	Pressure drop for Gas Flowrate in selected I.D. only
TPF-D-009	Liquid hold-up and flow pattern identification, superficial gas and oil velocities
TPF-D-010	Pressure drop for two phase flow system

3.2 Theoretical Formulae:

$$x = \sqrt{\Delta P_L / \Delta P_G}$$

$$\Delta P_{tp} = \Delta P_G \phi GTT^2 \qquad \text{(PSI)}$$

$$\lambda = \left[(\rho_G/.075)(\rho_L/62.3)\right]^{.5}$$

$$\psi = (73/\gamma)(\mu_L(62.3/\rho_L)^2)^{1/3}$$

ϕGTT:

$$\text{Froth/bubble flow} = \frac{14.2\ x^{.75}}{L^{.1}} \qquad \text{(PSI)}$$

$$\text{Plug flow} = \frac{27.315\ x^{.855}}{L^{.17}} \qquad \text{(PSI)}$$

$$\text{Stratified flow} = \frac{15{,}400\ x}{L^{.8}} \qquad \text{(PSI)}$$

$$\text{Slug flow} = \frac{1{,}190\ x^{.815}}{L^{.5}} \qquad \text{(PSI)}$$

$$^{*}\text{Annular flow} = (4.8 - .3125D)\ x^{.343} - .021D$$

*Substitute 10.00 for D in all pipelines with D greater than 10.00.

$$\text{Wave flow} = \frac{f_{tp}\ G^2\ L'}{193.2\ d\rho G}$$

$$\Delta P_{TPH} = \Delta P_{TP} + \frac{N\ H\ F_e\ \rho_L}{144}$$

$$F_e = k/v^{0.7} = .00967\ (L)^{.5}/V^{.7}\ , \quad V > 10\ \text{ft./sec.}$$

$$L = \frac{(B/D)(42)(\text{lb./gal.})}{(24)(A_f)}$$

$$G = \frac{(\text{mcf})(1000)(G)(\text{air MW})}{(24)(379)(A_f)}$$

Slug Volume Computation

$$\%\ \text{pipeline volume @ } Qb_1 = U_1 \qquad Qb_2 > Qb_1$$
$$\%\ \text{pipeline volume @ } Qb_2 = U_2$$

$$\left((U_1 - U_2) \times \frac{L'}{5{,}280} \times \text{LFV} \right) = \text{slug size}$$

Gas Density

$$\rho_G = \frac{(P_m)\ M}{(z)(10.73)(T)} \qquad (\text{lbs./ft.}^3)$$

Gas Viscosity

$$\mu_G = C \exp\left[x \left[\frac{\rho_G}{62.43}\right]^{y} \right] \qquad (\text{lb.m./ft.-sec.})$$

$$C = \frac{.001\ (\ (7.77 + .0063M)\ T^{1.5})}{122.4 + 12.9M + T}$$

$$x = 2.57 + \frac{1914.5}{T} + .0095M$$

$$y = 1.11 + .04x$$

$$T = {}^\circ R$$

$$R_s = \left[\frac{\dfrac{(G\ P_m)}{18}\ (10^{API/80})}{100.00091T} \right]^{1.205} \qquad (\text{ft.}^3/\text{BBL})$$

$$API = \frac{141.5}{SG_{oil}} - 131.5$$

Oil Formation Volume Factor: Standing Correlation

$$F = Rs\,(G/sg_{oil})^{.5} + 1.25T$$

$$B_o = .972 + .000147\,F^{1.175} \qquad (BBL/STB)$$

$$\rho_L = \frac{1}{B_o}\,(\rho_{oil_{sc}} + \rho_{Gsc}\,R_s/5.61) \qquad (lbs./ft.^3)$$

Saturated Oil Viscosity

$$\mu_{os} = A(\mu_{od})^B$$

$$A = .2 + \frac{0.8}{10^{.00081R_s}}, \qquad B = .43 + \frac{.57}{10^{.00072\,R_s}}$$

Oil Viscosity Approximation

$$\mu_{od} = 1.5\,(66.2\,(sg_{oil})^{17.24}) + 15\% \qquad (@100°F)$$

Two Phase Flow Computations

STEP	INSTRUCTIONS	KEYS	OUTPUT UNITS
1.	load card TPF-D-001, side 1 & 2		
2.	input flowrate - gas	A	mcfd
3.	input flowrate - oil	A	bpd
4.	input specific gravity - oil	A	n
5.	input speicifc gravity - gas	A	n
6.	input flowing temperature	A	^oF
7.	input inlet pressure	A	psia
8.	input outlet pressure	A	psia
9.	input barometric pressure, (or constant 0.00 when P_1 & P_2 in psia)	A	0.00
10.	input line length	A	miles
11.	Compute:	B	
	(1) average pressure		psia
	(2) liquid degrees API		n
	(3) gas solubility ratio - R_s		ft.^{3}/BBL
	(4) F factor		n
	(5) oil formation factor - B_o		BBL/STB
	(6) actual liquid volume		bpd
	(7) gas density at standard conditions		lb./ft.^{3}
	(8) liquid density at standard conditions		lb./ft.^{3}
	(9) saturated oil density		lb./ft.^{3}
	(10) saturated oil density		n (ratio)
12.	load card TPF-D-002, side 1		
13.	Compute:	A	
	(1) unsaturated oil viscosity		centipoise
	(2) B factor		n
	(3) A factor		n
	(4) saturated oil viscosity		centipoise
	(5) saturated oil viscosity		centistokes
	(6) pressure drop per mile		PSI/mi.
14.	load card TPF-D-003, side 1 & 2		
15.	Compute:	B	
	(1) critical pressure		psia
	(2) reduced pressure		n
	(3) critical temperature		^oR
	(4) reduced temperature		n
	(5) compressibility		n
	(6) gas density at operating conditions		lb./ft.^{3}
	(7) x factor		n
	(8) y factor		n
	(9) c factor		n
	(10) gas viscosity		centipoise
	(11) gas viscosity		lb.m./ft.sec.
16.	load card TPF-D-004, side 1 & 2, Newton Approx. Data Card		
17.	load card TPF-D-005, side 1 & 2		

STEP	INSTRUCTIONS	KEYS	OUTPUT UNITS
18.	Input:		
	(1) a constant "0.00"	A	0.00
	(2) pressure drop per mile	A	PSI/mi.
	(3) constant ".433"	A	.433
	(4) flowrate of oil	A	bpd
	(5) viscosity of oil - saturated	A	centistokes
	(6) specific gravity - saturated	A	n
19.	Compute:		
	(1) internal diameter for a given velocity (option)	D	inches
	(2) velocity and internal diameter for flow conditions	E	ft./sec. inches
20.	load card TPF-D-006A, side 1 & 2		
21.	Clear all registers:	f CL REG f P$\gtrless$S f CL REG CL x	 0.00
	Input:		
	(1) flowrate for gas phase	A	mcfd
	(2) constant	A	1.0
	(3) upstream pressure	A	psia
	(4) downstream pressure	A	psia
	(5) (constant) barometric pressure	A	psia
	(6) specific gravity	A	n
	(7) line length	A	miles
	(8) flowing temperature	A	^oF
	(9) base pressure	A	psia
	(10) base temperature	A	^oF
	(11) viscosity	A	lb.m./ft.sec.
	(12) effective roughness	A	micro inches
	(13) drag factor	A	n
	(14) elevation difference	A	feet
22.	Compute:	B	
	(1) average pressure		psig
	(2) critical pressure		psia
	(3) critical temperature		^oR
	(4) reduced pressure		n
	(5) reduced temperature		n
	(6) compressibility		n
	(7) α factor		n
23.	load card TPF-006B, side 1 & 2		
24.	Compute:		
	(1) Weymouth approximation of internal diameter for gas phase	A	inches
	(2) Newton approximation of internal diameter for gas phase	E	inches

STEP	INSTRUCTIONS	KEYS	OUTPUT UNITS
	(3) Reynolds No. at flowing conditions	RCL D	n
	(4) Reynolds No. at transition from partially turbulent to fully turbulent flow.	RCL E	n
	(5) transmission factor $\sqrt{1/f_{ft}}$	RCL A	n
	(6) transmission factor $\sqrt{1/f_{pt}}$	f P≷S	n
		RCL 0	
		f P≷S	
25.	load card TPF-D-007, side 1 & 2		
26.	Clear all registers:	f CL REG	
		f P≷S	
		f CL REG	
		CL x	0.00
	Input:		
	(1) specific gravity of saturated oil	A	n
	(2) flowrate of oil	A	bpd
	(3) internal diameter for oil phase	A	inches
	(4) viscosity of saturated oil	A	centistokes
	(5) length of line	A	feet
27.	Compute:	B	
	(1) velocity of flow		ft./sec.
	(2) Reynolds No. modified		n
	(3) Reynolds No. critical		n
	(4) type of flow, where 1 = viscous 2 = intermediate 3 = partially turbulent 4 = fully turbulent		n
	(5) pressure drop per L		PSI/L
28.	load card TPF-D-008A, side 1 & 2		
29.	Clear all registers:	f CL REG	
		f P≷S	
		f CL REG	
		CL x	0.00
	Input:		
	(1) flowrate of gas	A	mcfd
	(2) viscosity of gas phase	A	lb.m./ft.sec.
	(3) gas specific gravity	A	n
	(4) selected outside diameter	A	inches
	(5) wall thickness	A	inches
	(6) base pressure	A	psia
	(7) base temperature	A	^oF
	(8) effective pipe roughness	A	micro inches
	(9) drag factor	A	n
	(10) elevation difference	A	feet
	(11) line length	A	miles

STEP	INSTRUCTIONS	KEYS	OUTPUT UNITS
	(12) flowing temperature	A	^oF
	(13) upstream pressure	A	psig
	(14) barometric pressure (constant = 0.00)	A	0.00
30.	Compute:	C	
	(1) Reynolds No.		n
	(2) transmission factor $\sqrt{1/f_{ft}}$		n
	(3) critical Reynolds No.		n
	(4) α factor		n
31.	load card TPF-D-008B, side 1 & 2		
32.	Compute:	B	
	(1) Weymouth approximation of downstream pressure		psig
	(2) Newton approximation of downstream pressure		psig
33.	(1) Recall P_1 from previous computations.	P_1 ENTER↑	psia
	(2) Recall P_2 from previous computations.	P_2 ENTER↑	psia
	(3) subtract ($P_1 - P_2$)	-	psia
34.	load card TPF-D-009, side 1 & 2		
35.	Clear all registers:	f CL REG	
		f P≷S	
		f CL REG	
		CL x	0.00
	Input:		
	(1) total pressure drop for oil phase in selected internal diameter	A	PSI
	(2) total pressure drop for gas phase in selected internal diameter	A	PSI
	(3) oil flowrate saturated	A	bpd
	(4) specific gravity of saturated oil	A	n
	(5) outside pipe diameter	A	inches
	(6) wall thickness	A	inches
	(7) gas flowrate	A	mcfd
	(8) specific gravity of gas	A	n
	(9) density of gas at operating conditions	A	$lb./ft.^3$
	(10) density of oil at operating conditions	A	$lb./ft.^3$
	(11) specific gravity of unsaturated oil	A	n
	(12) average pressure	A	psia
	(13) viscosity of oil saturated	A	centipoise

STEP	INSTRUCTIONS	KEYS	OUTPUT UNITS
36.	Compute:	B	
	(1) x factor		n
	(2) percentage pipe fill		%
	(3) mass velocity of liquid (L)		lb./hr.ft.^{2}
	(4) mass velocity of gas (G)		lb./hr.ft.^{2}
	(5) λ factor		n
	(6) surface tension factor		n
	(7) surface tension		dynes/cm.
	(8) ψ factor		n
	(9) G/λ factor		n
	(10) $\psi L\lambda/G$ factor		n
	(11) gas superficial velocity		ft./sec.
	(12) liquid superficial velocity		ft./sec.
37.	load card TPF-D-010, side 1 & 2		
38.	(1) recall storage register A (ordinate)	RCL A	n
	(2) recall storage register B (abscissa)	RCL B	n
	(3) locate point on two phase flow regime graph		
	(4) identify flow pattern from the graph. ie. slug, etc.		
39.	Input:		
	(1) internal diameter, ENTER↑	E	inches
	(2) length of line, ENTER↑	R/S	feet
	(3) no. of hills, ENTER↑	R/S	n
	(4) average elevation of hills, ENTER↑	R/S	feet
	(5) gas viscosity, ENTER↑	R/S	centipoise
40.	Enter number for pattern flow, ie. 1 = slug 2 = stratified 3 = plug 4 = froth/bubble 5 = annular mist 6 = wave	ENTER↑	n
41.	Compute pressure drop for flow pattern.	A	PSI
42.	Compute pressure drop for elevation factors.	B	PSI
43.	Total pressure drop: Pattern flow pressure drop + Elevation pressure drop = Total pressure drop		

Example: TPF-D-001 through TPF-D-010

Computation Procedure

Given Conditions:

length of pipeline: 134,370 feet
inlet pressure: 424.7 psia
outlet pressure: 319.7 psia
volumetric flowrate of stock tank oil: 6,912 bpd
volumetric flowrate of gas: 3,191 mcfd
specific gravity of gas: .7
flowing temperature: 60°F
specific gravity of liquid: .84
total sum of elevations: 124.1 feet
constant - barometric pressure: 0.00

Execution: Load card TPF-D-001, side 1 & 2

Store parameters in this order:

KEYSTROKES	OUTPUT
3,191 "A"	3,191.00
6,912 "A"	6,912.00
.84 "A"	.84
.70 "A"	.70
60 "A"	60.00
424.7 "A"	424.70
319.7 "A"	319.70
0 "A"	0.00
25.45 "A"	25.45

To Compute:

Press "B"	"376.54" psia (average pressure)
	"36.95" °API (oil API gravity)
	"78.60" ft.^{3}/BBL (R_s factor)
	"2,647.75" mscfd (actual gas volume)
	"146.75" (F factor)
	"1.02" BBL/STB (B_o)
	"7,075.44" bpd (actual oil volume)
	".05" lb./ft.^{3} (gas density at standard conditions)
	"52.42"lb./ft.^{3} (oil density at standard conditions - dead oil)
	"51.94" lb./ft.^{3} (density of saturated oil)
	".83" (density of saturated oil - ratio)

Load card TPF-D-002, side 1

KEYSTROKES	OUTPUT
Press "A"	"4.91" centipoise (dead oil viscosity)
	".93" (B factor)
	".89" (A factor)
	"3.92" centipoise (saturated oil viscosity)
	"4.71" centistokes (saturated oil viscosity)
	"4.13" PSI/mi. (pressure drop per mile for given P_1, P_2 & L)

Load card TPF-D-003, side 1 & 2

Press "B"	"666.40" psia (P_c)
	".57" (P_r)
	"390.00" °R (T_c)
	"1.33" (T_r)
	".92" (z)
	"1.49" lb./ft.^{3} (ρ_{gas} at operating conditions)
	"6.44" (x factor)
	"1.37" (y factor)
	".01" (c factor)
	".01" centipoise (μ_g)
	"7.24 x 10^{-6}" lb.m./ft.sec. (μ_g)

Load card TPF-D-004, side 1 & 2, Newton Approx. Data Card

Load card TPF-D-005, side 1 & 2 - Internal Diameter Computation for liquid component

Store parameters in this order:

0.00 "A"	0.000
4.13 "A"	4.130
.433 "A"	.433
7,075.44 "A"	7,075.440
4.71 "A"	4.710
.83 "A"	.830

To Compute:

7.00 "D"	3.46818 inches (diameter approximation)

Press "E"	"1.76" ft./sec. (velocity)
	"6.91597" inches (Newton approximation)

Load card TPF-D-006A - Gas component, Internal Diameter Computation; Clear all Registers.

f CL REG	
f P≷S	
f CL REG	
CL x	0.00

Store parameters in this order:

KEYSTROKES	OUTPUT
2,647.75 "A"	2,647.75
1.00 "A"	1.00
424.7 "A"	424.70
319.7 "A"	319.70
0 "A"	0.00
.7 "A"	.70
25.45 "A"	25.45
60 "A"	60.00
14.73 "A"	14.73
60 "A"	60.00
7.24×10^{-6} "A"	7.24×10^{-6}
750 "A"	750.00
.98 "A"	.98
0 "A"	0.00

To Compute:

Press "B"	"376.54" psia (average pressure)
	"666.40" psia (P_c)
	"390.00" ^oR (T_c)
	".57" (P_r)
	"1.33" (T_r)
	".92" (z)
	"3,462,655.05" (α)

Load card TPF-D-006B

Press "A"	4.49 (Weymouth approximation for internal diameter)
Press "E"	4.24 (Newton approximation for internal diameter)
RCL D	815,454.68 (Reynolds No.)
RCL E	621,046.26 (critical Reynolds No.)
RCL A	17.28 ($\sqrt{1/f_{ft}}$

Note: Secondary Register, f P≷S; RCL 0 when flow is partially turbulent.

The Newton Approximation of internal diameter for the oil component from previous calculations = 6.91 inches.

The Newton Approximation of internal diameter for the gas component from the previous calculations = 4.24 inches.

Add both Newton Approximations = 11.15 inches, therefore the internal diameter to accommodate the two phase flow volume is a 12.00 inch internal diameter line. - 1st guess.

Load card TPF-D-007, side 1 & 2: CLEAR ALL REGISTERS

Store parameters in this order:

KEYSTROKES	OUTPUT
.83 "A"	.830
7,075.64 x 1.1 "A" (90% efficiency)	7,783.204
12 "A"	12.000
4.71 "A"	4.710
134,370 "A"	134,370.000

To Compute:

Press "B" ________ ".643" ft./sec. (velocity)
________ "1.64" (R^*)
________ "353.99" (R^* critical)
________ "3.0" (type of flow)
________ "9.13" PSI (pressure drop per L)

Load card TPF-D-008A, side 1 & 2: CLEAR ALL REGISTERS

Store parameters in this order:

2647.75 x 1.1 "A"	2,912.525
7.24×10^{-6} "A"	7.24×10^{-6}
.7 "A"	.700
12.750 "A"	12.750
.375 "A"	.375
14.73 "A"	14.730
60 "A"	60.000
750 "A"	750.000
.98 "A"	.980
0.00 "A"	0.000
25.45 "A"	25.450
60 "A"	60.000
424.7 "A"	424.700
0 "A"	0.000

To Compute:

Press "C" ________ "317,410.047" (Reynolds No.)
________ "19.089" ($\sqrt{1/f_{ft}}$)
________ "1,979,856.735" (critical Reynolds No.)
________ "638.503" (α factor)

Load card TPF-D-008B, side 1 & 2

To Compute:

Press "B" ________ "424.12" psia (Weymouth approximation of downstream pressure)
________ "424.02" psia (Newton approximation of downstream pressure)

The inlet pressure from previous calculations = 424.7 psia.

The Newton approximation of outlet pressure from the above calculation = 424.12 psia.

Subtract 424.7 - 424.02 = <u>0.68 psia</u>.

Load card TPF-D-009: CLEAR ALL REGISTERS

Store parameters in this order:

KEYSTROKES	OUTPUT
9.134593 "A"	9.134593
.68 "A"	.680000
7075.64 x 1.1 "A"	7,783.204000
.83 "A"	.830000
12.750 "A"	12.750000
.375 "A"	.375000
2647.75 x 1.1 "A"	2,912.525000
.7 "A"	.700000
1.49 "A"	1.490000
51.94 "A"	51.940000
.84 "A"	.840000
376.84 "A"	376.840000
3.92 "A"	3.920000

To Compute:

Press "B"	"3.66" (x factor)
	"46.26" % (percentage linefill)
	"119,902.8966" lb./hr.ft.^{2} (L)
	"8,276.09308" lb./hr.ft.^{2} (G)
	"4.06977" (λ)
	".75256" (surface tension factor)
	"28.85760" dynes/cm. (surface tension factpr)
	"6.05271" (ψ factor)
	"2,033.55496" (G/λ factor)
	"356.88132" ($\psi L\lambda/G$ factor)
	"1.54290" ft./sec. (V_g)
	".64125" ft./sec. (V_L)

Load card TPF-D-010, side 1 & 2

RCL A ________ 2033.55 (flow pattern graph ordinate)

RCL B ________ 356.88 (flow pattern graph abscissa)

Input the following as shown:

12.0 ENTER↑ E ________ 12.0 (internal diameter)
134,370 ENTER↑ R/S ________ 134,370 (line length)

KEYSTROKES	OUTPUT
1.00 ENTER↑ R/S	1.00 (number of hills)
124.00 ENTER↑ R/S	124.00 (average elevation - feet)
.01 ENTER↑ R/S	.01 (gas viscosity - cp.)

Go to flow pattern graph and locate point (356.88, 2033.55).
Pattern appears to be slug flow.

1.00 ENTER↑ A________66.69 PSI (pressure drop for slug flow)

Press "B"___________52.83 PSI (elevation pressure drop)

Add the two pressure drops: 66.69 + 52.83 = 119.52 PSI.
This represents the computed pressure drop for the two phase flow system.

To Compute the % error from initial conditions:

119.52 - (424.7 - 319.7) = 14.52 PSI

100 (14.52 ÷ 105) = 14%

Since the maximum allowable range is ± 34% of the initial conditions, then 119.52 PSI total pressure drop is acceptable with only 14% error.

Note: If the computed error had been greater than 34% then the entire procedure must be repeated with a different internal diameter.

FLOW PATTERN GRAPH

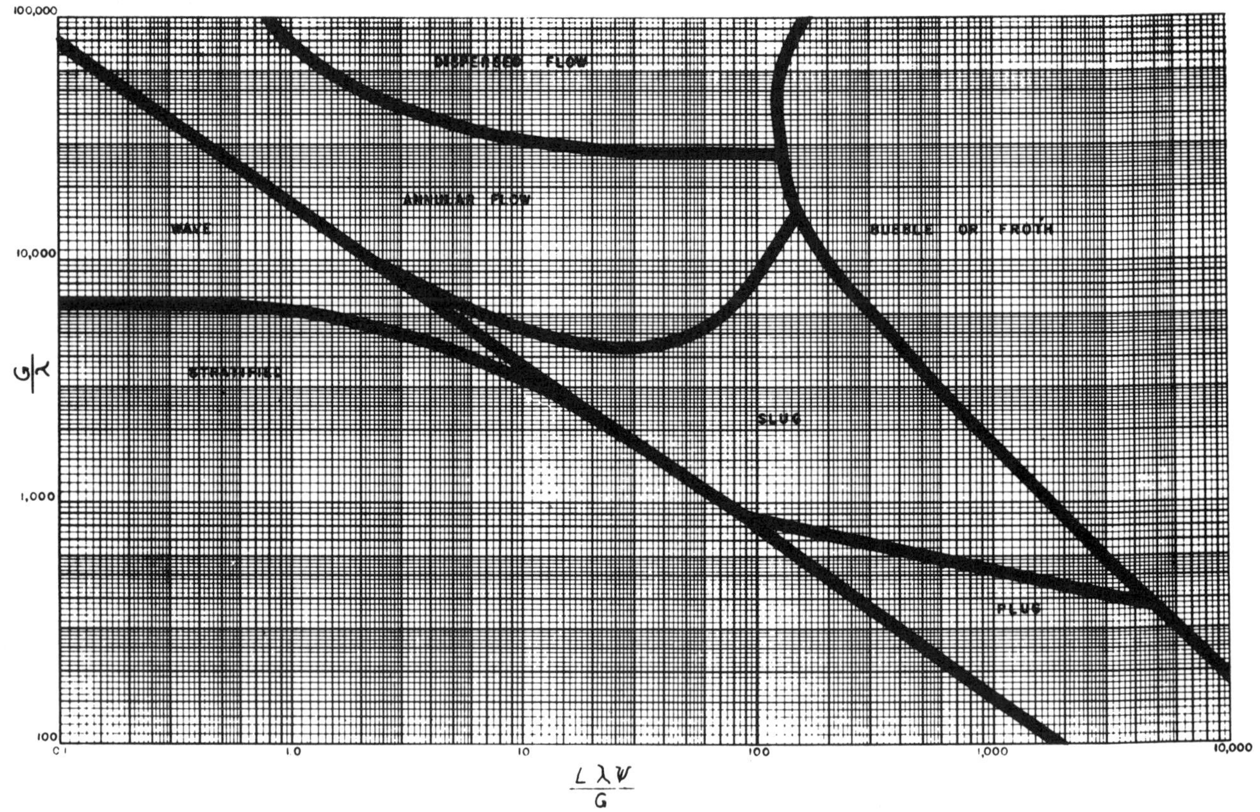

TPF-D-001

STEP	KEY ENTRY	KEY CODE
001	f LBL A	31 25 11
	STO (i)	33 24
	f GSB 1	31 22 01
	h RTN	35 22
	f LBL 1	31 25 01
	f ISZ	31 34
	R/S	84
	h ST I	35 33
	GTO 1	22 01
010	h RTN	35 22
	f LBL B	31 25 12
	RCL 5	34 05
	RCL 7	34 07
	+	61
	3	03
	h y^x	35 63
	RCL 6	34 06
	RCL 7	34 07
	+	61
020	3	03
	h y^x	35 63
	−	51
	STO (i)	33 34
	RCL 5	34 05
	RCL 7	34 07
	+	61
	g x^2	32 54
	RCL 6	34 06
	RCL 7	34 07
030	+	61
	g x^2	32 54
	−	51
	RCL (i)	34 24
	h $x \gtrless y$	35 52
	÷	81
	.	83
	6	06
	7	07
	x	71
040	STO A	33 11
	f -x-	31 84
	1	01
	4	04
	1	01
	.	83
	5	05
	RCL 2	34 02
	÷	81
	1	01
050	3	03
	1	01
	.	83
	5	05
	−	51
	STO (i)	33 24
	STO B	33 12
	f -x-	31 84
	8	08
	0	00
060	÷	81
	g 10^x	32 53
	RCL 3	34 03
	x	71
	RCL A	34 11
	x	71
	1	01
	8	08
	÷	81
	STO (i)	33 24
070	RCL 4	34 04
	.	83
	0	00
	0	00
	0	00
	9	09
	1	01
	x	71
	g 10^x	32 53
	RCL (i)	34 24
080	h $x \gtrless y$	35 52
	÷	81
	1	01
	.	83
	2	02
	0	00
	5	05
	h y^x	35 63
	STO C	33 13
	f -x-	31 84
090	RCL 1	34 01
	x	71
	1	01
	0	00
	0	00
	0	00
	÷	81
	RCL 0	34 00
	h $x \gtrless y$	35 52
	−	51
100	STO D	33 14
	f -x-	31 84
	RCL 3	34 03
	RCL 2	34 02
	÷	81
	f $\sqrt{x}$	31 54
	RCL C	34 13
	x	71
	1	01
	.	83
110	2	02
	5	05
	RCL 4	34 04

STEP	KEY ENTRY	KEY CODE
	x	71
	+	61
	f -x-	31 84
	1	01
	.	83
	1	01
	7	07
120	5	05
	h y^x	35 63
	.	83
	0	00
	0	00
	0	00
	1	01
	4	04
	7	07
	x	71
130	.	83
	9	09
	7	07
	2	02
	+	61
	STO (i)	33 24
	f -x-	31 84
	RCL 1	34 01
	x	71
	STO E	33 15
140	f -x-	31 84
	1	01
	4	04
	.	83
	7	07
	3	03
	RCL 3	34 03
	x	71
	2	02
	9	09
150	x	71
	1	01
	0	00
	.	83
	7	07
	3	03
	÷	81
	RCL 4	34 04
	4	04
	6	06
160	0	00
	+	61
	÷	81
	f -x-	31 84
	RCL C	34 13
	x	71
	5	05
	.	83
	6	06
	1	01
170	÷	81
	RCL 2	34 02
	6	06
	2	02
	.	83
	4	04
	x	71
	f -x-	31 84
	+	61
	RCL (i)	34 24
180	÷	81
	f -x-	31 84
	6	06
	2	02
	.	83
	4	04
	÷	81
	STO 9	33 09
	h RTN	35 22
	g LBLf b	32 25 12
190	STO 5	33 05
	h RTN	35 22
	g LBLf c	32 25 13
	STO 6	33 06
	h RTN	35 22
	g LBLf d	32 25 14
	STO 1	33 01
	h RTN	35 22
	g LBLf e	32 25 15
	STO 1	33 01
200	h RTN	35 22
	g LBLf a	32 25 11
	f LOG	31 43
	f $P \gtrless S$	31 42
	f LOG	31 43
	f $P \gtrless S$	31 42
	h RTN	35 22

STEP	KEY ENTRY	KEY CODE
001	f LBL A	31 25 11
	RCL 2	34 02
	f P≷S	31 42
	1	01
	7	07
	.	83
	2	02
	4	04
	h y^x	35 63
010	6	06
	6	06
	.	83
	1	01
	7	07
	x	71
	1	01
	.	83
	5	05
	x	71
020	STO 0	33 00
	f -x-	31 84
	RCL C	34 13
	.	83
	0	00
	0	00
	0	00
	7	07
	2	02
	x	71
030	g 10^x	32 53
	.	83
	5	05
	7	07
	h x≷y	35 52
	÷	81
	.	83
	4	04
	3	03
	+	61
040	STO 1	33 01
	f -x-	31 84
	RCL C	34 13
	.	83
	0	00
	0	00
	0	00
	8	08
	1	01
	x	71
050	g 10^x	32 53
	.	83
	8	08
	h x≷y	35 52
	÷	81
	.	83
	2	02
	+	61
	STO 2	33 02
	f -x-	31 84
060	RCL 0	34 00
	RCL 1	34 01
	h y^x	35 63
	RCL 2	34 02
	x	71
	STO 3	33 03
	f -x-	31 84
	f P≷S	31 42
	RCL 9	34 09
	÷	81
070	f -x-	31 84
	0	00
	h ST I	35 33
	h R ↓	35 53
	RCL 5	34 05
	RCL 6	34 06
	−	51
	RCL 8	34 08
	÷	81
	f -x-	31 84
080	h RTN	35 22

TPF-D-003

STEP	KEY ENTRY	KEY CODE
001	f LBL B	31 25 12
	RCL 3	34 03
	4	04
	3	03
	CHS	42
	x	71
	6	06
	9	09
	6	06
010	.	83
	5	05
	+	61
	f -x-	31 84
	RCL A	34 11
	h x≷y	35 52
	÷	81
	f P≷S	31 42
	STO 4	33 04
	f P≷S	31 42
020	f -x-	31 84
	RCL 3	34 03
	3	03
	0	00
	0	00
	x	71
	1	01
	8	08
	0	00
	+	61
030	f -x-	31 84
	RCL 4	34 04
	4	04
	6	06
	0	00
	+	61
	h x≷y	35 52
	÷	81
	f P≷S	31 42
	STO 5	33 05
040	f P≷S	31 42
	f -x-	31 84
	f P≷S	31 42
	RCL 5	34 05
	.	83
	3	03
	4	04
	x	71
	.	83
	6	06
050	CHS	42
	+	61
	RCL 4	34 04
	x	71
	1	01
	+	61
	STO 6	33 06
	f -x-	31 84
	f P≷S	31 42
	RCL 3	34 03
060	2	02
	9	09
	x	71
	RCL A	34 11
	x	71
	STO B	33 12
	RCL 4	34 04
	4	04
	6	06
	0	00
070	+	61
	1	01
	0	00
	.	83
	7	07
	3	03
	x	71
	f P≷S	31 42
	RCL 6	34 06
	f P≷S	31 42
080	x	71
	RCL B	34 12
	h x≷y	35 52
	÷	81
	f P≷S	31 42
	STO 7	33 07
	f P≷S	31 42
	f -x-	31 84
	RCL 3	34 03
	2	02
090	9	09
	x	71
	.	83
	0	00
	0	00
	9	09
	5	05
	x	71
	2	02
	.	83
100	5	05
	7	07
	+	61
	STO B	33 12
	RCL 4	34 04
	4	04
	6	06
	0	00
	+	61
	1	01
110	9	09
	1	01
	4	04

STEP	KEY ENTRY	KEY CODE
	.	83
	5	05
	h x≷y	35 52
	÷	81
	RCL B	34 12
	+	61
	f P≷S	31 42
120	STO 8	33 08
	f P≷S	31 42
	f -x-	31 84
	f P≷S	31 42
	RCL 8	34 08
	.	83
	0	00
	4	04
	x	71
	1	01
130	.	83
	1	01
	1	01
	+	61
	STO 9	33 09
	f P≷S	31 42
	f -x-	31 84
	RCL 3	34 03
	2	02
	9	09
140	x	71
	.	83
	0	00
	0	00
	6	06
	3	03
	x	71
	7	07
	.	83
	7	07
150	7	07
	+	61
	RCL 4	34 04
	4	04
	6	06
	0	00
	+	61
	1	01
	.	83
	5	05
160	h y^x	35 63
	x	71
	.	83
	0	00
	0	00
	0	00
	1	01
	x	71
	STO B	33 12

STEP	KEY ENTRY	KEY CODE
	RCL 3	34 03
170	2	02
	9	09
	x	71
	1	01
	2	02
	.	83
	9	09
	x	71
	RCL 4	34 04
	4	04
180	6	06
	0	00
	+	61
	+	61
	1	01
	2	02
	2	02
	.	83
	4	04
	+	61
190	RCL B	34 12
	h x≷y	35 52
	÷	81
	STO B	33 12
	f -x-	31 84
	f P≷S	31 42
	RCL 7	34 07
	6	06
	2	02
	.	83
200	4	04
	3	03
	÷	81
	RCL 9	34 09
	h y^x	35 63
	RCL 8	34 08
	x	71
	g e^x	32 52
	RCL B	34 12
	x	71
210	STO B	33 12
	f -x-	31 84
	6	06
	.	83
	7	07
	2	02
	EEX	43
	CHS	42
	4	04
	x	71
220	f P≷S	31 42
	h RTN	35 22

TPF-D-004 - Newton Approximation Data Card

Card TPF-D-004 contains parameters in storage and is defined as a data card.

Storage Register No.	
2	.433
6	.0119
7	.241
8	.135
9	.00207
S0	-.3192
S1	.008140
S2	.355
S3	.00662
S4	.0018
S5	USED
S6	1.05
S7	3.7
S8*	.000750
S9	USED

* This register is used for the roughness factor and may be varied by direct input if required.

STEP	KEY ENTRY	KEY CODE
001	f LBL A	31 25 11
	STO (i)	33 24
	f GSB 3	31 22 03
	h RTN	35 22
	f LBL 3	31 25 03
	f ISZ	31 34
	R/S	84
	h ST I	35 33
	GTO 3	22 03
010	h RTN	35 22
	f LBL E	31 25 15
	f FIX	31 23
	GTO B	22 12
	g LBLf a	32 25 11
	RCL B	34 12
	GTO 0	22 00
	f LBL 6	31 25 06
	RCL 0	34 00
	f GSB 1	31 22 01
020	STO B	33 12
	f LBL 0	31 25 00
	RCL A	34 11
	RCL 0	34 00
	STO A	33 11
	-	51
	RCL D	34 14
	RCL B	34 12
	STO D	33 14
	-	51
030	÷	81
	x	71
	STO -0	33 51 00
	RCL 0	34 00
	h PAUSE	35 72
	÷	81
	f RND	31 24
	f x≠0	31 61
	GTO 6	22 06
	g GSB e	32 22 15
040	f -x-	31 84
	RCL 0	34 00
	h RTN	35 22
	f LBL B	31 25 12
	.	83
	0	00
	1	01
	f %	31 82
	STO C	33 13
	2	02
050	÷	81
	-	51
	STO A	33 11
	STO 0	33 00
	f GSB 1	31 22 01
	STO D	33 14
	RCL A	34 11
	RCL C	34 13
	+	61
	STO 0	33 00
060	f GSB 1	31 22 01
	STO B	33 12
	RCL D	34 14
	-	51
	RCL C	34 13
	÷	81
	GTO f a	22 31 11
	f LBL 1	31 25 01
	RCL 9	34 09
	.	83
070	0	00
	0	00
	2	02
	0	00
	7	07
	-	51
	f x=0	31 51
	GTO 2	22 02
	7	07
	ENTER ↑	41
080	0	00
	÷	81
	h RTN	35 22
	f LBL 2	31 25 02
	RCL 7	34 07
	RCL 3	34 03
	g x^2	32 54
	x	71
	RCL 5	34 05
	x	71
090	RCL 1	34 01
	÷	81
	.	83
	2	02
	h y^x	35 63
	f P⪋S	31 42
	STO 5	33 05
	f GSB 9	31 22 09
	.	83
	2	02
100	h y^x	35 63
	RCL 5	34 05
	x	71
	f P⪋S	31 42
	RCL 0	34 00
	-	51
	h RTN	35 22
	f LBL 9	31 25 09
	f P⪋S	31 42
	RCL 6	34 06
110	RCL 3	34 03
	x	71
	RCL 0	34 00

STEP	KEY ENTRY	KEY CODE
	÷	81
	RCL 4	34 04
	÷	81
	f P≷S	31 42
	STO 9	33 09
	.	83
	4	04
120	g x≤y	32 71
	GTO 5	22 05
	RCL 9	34 09
	f P≷S	31 42
	RCL 8	34 08
	g x≤y	32 71
	GTO 7	22 07
	h x≷y	35 52
	h 1/x	35 62
	RCL 9	34 09
130	x	71
	f P≷S	31 42
	h RTN	35 22
	f LBL 5	31 25 05
	RCL 9	34 09
	GTO 8	22 08
	g LBLf d	32 25 14
	g x≤y	32 71
	GTO f c	22 31 13
	h x≷y	35 52
140	h 1/x	35 62
	f P≷S	31 42
	RCL 2	34 02
	h y^x	35 63
	RCL 3	34 03
	x	71
	RCL 4	34 04
	+	61
	h RTN	35 22
	f LBL 8	31 25 08
150	f P≷S	31 42
	RCL 0	34 00
	f P≷S	31 42
	RCL 8	34 08
	÷	81
	RCL 7	34 07
	x	71
	STO 9	33 09
	f LOG	31 53
	4	04
160	x	71
	RCL 9	34 09
	RCL 6	34 06
	h y^x	35 63
	x	71
	1	01
	.	83
	4	04
	x	71
	7	07
170	7	07
	4	04
	2	02
	÷	81
	f P≷S	31 42
	GTO f d	22 31 14
	h RTN	35 22
	f LBL 7	31 25 07
	h x≷y	35 52
	f P≷S	31 42
180	RCL 0	34 00
	h y^x	35 63
	RCL 1	34 01
	x	71
	h RTN	35 22
	g LBLf c	32 25 13
	f P≷S	31 42
	RCL 9	34 09
	f LOG	31 53
	4	04
190	x	71
	g x^2	32 54
	h 1/x	35 62
	f -x-	31 84
	h RTN	35 22
	g LBLf b	32 25 12
	-	51
	RCL 5	34 05
	RCL 2	34 02
	x	71
200	x	71
	RCL 0	34 00
	÷	81
	STO +1	33 61 01
	h RTN	35 22
	g LBLf e	32 25 15
	RCL 6	34 06
	RCL 3	34 03
	x	71
	RCL 0	34 00
210	g x^2	32 54
	÷	81
	h RTN	35 22
	f LBL C	31 25 13
	0	00
	h ST I	35 33
	h RTN	35 22
	f LBL D	31 25 14
	h 1/x	35 62
	RCL 3	34 03
220	RCL 6	34 06
	x	71
	x	71
	f $\sqrt{x}$	31 54
	h RTN	35 22

STEP	KEY ENTRY	KEY CODE
001	f LBL A	31 25 11
	STO (i)	33 24
	f GSB 1	31 22 01
	h RTN	35 22
	f LBL 1	31 25 01
	f ISZ	31 34
	R/S	84
	h ST I	35 33
	GTO 1	22 01
010	h RTN	35 22
	f LBL B	31 25 12
	RCL 2	34 02
	RCL 4	34 04
	+	61
	3	03
	h y^x	35 63
	RCL 3	34 03
	RCL 4	34 04
	+	61
020	3	03
	h y^x	35 63
	-	51
	STO (i)	33 24
	RCL 2	34 02
	RCL 4	34 04
	+	61
	g x^2	32 54
	RCL 3	34 03
	RCL 4	34 04
030	+	61
	g x^2	32 54
	-	51
	RCL (i)	34 24
	h $x \gtrless y$	35 52
	STO (i)	33 24
	÷	81
	.	83
	6	06
	7	07
040	x	71
	RCL 4	34 04
	-	51
	f -x-	31 84
	STO A	33 11
	RCL 5	34 05
	4	04
	3	03
	x	71
	CHS	42
050	6	06
	9	09
	6	06
	.	83
	5	05
	+	61
	STO B	33 12
	f -x-	31 84
	RCL 5	34 05
	3	03
060	0	00
	0	00
	x	71
	1	01
	8	08
	0	00
	+	61
	STO C	33 13
	f -x-	31 84
	RCL A	34 11
070	RCL 4	34 04
	+	61
	RCL B	34 12
	÷	81
	f -x-	31 84
	STO D	33 14
	RCL 7	34 07
	4	04
	6	06
	0	00
080	+	61
	RCL C	34 13
	÷	81
	f -x-	31 84
	STO E	33 15
	1	01
	.	83
	2	02
	5	05
	g $x \leq y$	32 71
090	GTO 2	22 02
	RCL E	34 15
	.	83
	9	09
	x	71
	1	01
	.	83
	3	03
	-	51
	RCL D	34 14
100	x	71
	1	01
	+	61
	f -x-	31 84
	STO B	33 12
	GTO 3	22 03
	f LBL 2	31 25 02
	RCL E	34 15
	.	83
	3	03
110	4	04
	x	71
	.	83

STEP	KEY ENTRY	KEY CODE
	6	06
	-	51
	RCL D	34 14
	x	71
	1	01
	+	61
	STO B	33 12
120	f -x-	31 84
	GTO 3	22 03
	f LBL 3	31 25 03
	RCL (i)	34 24
	f GSB 4	31 22 04
	-	51
	RCL 5	34 05
	÷	81
	RCL 6	34 06
	÷	81
130	RCL 7	34 07
	4	04
	6	06
	0	00
	+	61
	÷	81
	RCL B	34 12
	÷	81
	f $\sqrt{x}$	31 54
	GTO 5	22 05
140	f LBL 4	31 25 04
	RCL A	34 11
	g x^2	32 54
	RCL 5	34 05
	x	71
	f P≷S	31 42
	RCL 3	34 03
	x	71
	f P≷S	31 42
	RCL 7	34 07
150	4	04
	6	06
	0	00
	+	61
	÷	81
	RCL B	34 12
	÷	81
	.	83
	0	00
	3	03
160	7	07
	5	05
	x	71
	h RTN	35 22
	f LBL 5	31 25 05
	3	03
	8	08
	.	83
	7	07

STEP	KEY ENTRY	KEY CODE
	7	07
170	4	04
	x	71
	RCL 9	34 09
	4	04
	6	06
	0	00
	+	61
	RCL 8	34 08
	÷	81
	x	71
180	1	01
	0	00
	0	00
	0	00
	÷	81
	RCL 0	34 00
	h x≷y	35 52
	÷	81
	f P≷S	31 42
	STO 8	33 08
190	RCL 0	34 00
	f P≷S	31 42
	RCL 9	34 09
	4	04
	6	06
	0	00
	+	61
	x	71
	h 1/x	35 62
	RCL 5	34 05
200	x	71
	RCL 8	34 08
	x	71
	RCL 0	34 00
	x	71
	.	83
	4	04
	7	07
	7	07
	5	05
210	x	71
	f P≷S	31 42
	STO 9	33 09
	f P≷S	31 42
	h RTN	35 22

STEP	KEY ENTRY	KEY CODE
001	f LBL E	31 25 15
	f FIX	31 23
	GTO B	22 12
	g LBLf a	32 25 11
	RCL 2	34 02
	GTO 0	22 00
	f LBL 6	31 25 06
	RCL 0	34 00
	f GSB 9	31 22 09
010	STO 2	33 02
	f LBL 0	31 25 00
	RCL 1	34 01
	RCL 0	34 00
	STO 1	33 01
	-	51
	RCL 4	34 04
	RCL 2	34 02
	STO 4	33 04
	-	51
020	÷	81
	x	71
	STO -0	33 51 00
	RCL 0	34 00
	h PAUSE	35 72
	÷	81
	f RND	31 24
	f x≠0	31 61
	GTO 6	22 06
	RCL 0	34 00
030	h RTN	35 22
	f LBL B	31 25 12
	.	83
	0	00
	1	01
	f %	31 82
	STO 3	33 03
	2	02
	÷	81
	-	51
040	STO 1	33 01
	STO 0	33 00
	f GSB 9	31 22 09
	STO 4	33 04
	RCL 1	34 01
	RCL 3	34 03
	+	61
	STO 0	33 00
	f GSB 9	31 22 09
	STO 2	33 02
050	RCL 4	34 04
	-	51
	RCL 3	34 03
	÷	81
	GTO f a	22 31 11
	f LBL 9	31 25 09
	RCL 0	34 00

STEP	KEY ENTRY	KEY CODE
	f P≷S	31 42
	STO 5	33 05
	f P≷S	31 42
060	h ST I	35 33
	GTO 1	22 01
	f LBL 3	31 25 03
	h RC I	35 34
	2	02
	.	83
	5	05
	h y^x	35 63
	x	71
	RCL 8	34 08
070	-	51
	f P≷S	31 42
	h RTN	35 22
	f LBL 1	31 25 01
	RCL 0	34 00
	f P≷S	31 42
	RCL 9	34 09
	h x≷y	35 52
	÷	81
	STO D	33 14
080	RCL 5	34 05
	RCL B	34 12
	÷	81
	3	03
	.	83
	7	07
	x	71
	STO 6	33 06
	f LOG	31 53
	4	04
090	x	71
	STO A	33 11
	RCL 6	34 06
	RCL C	34 13
	h 1/x	35 62
	h y^x	35 63
	1	01
	.	83
	4	04
	x	71
100	RCL A	34 11
	x	71
	STO E	33 15
	RCL D	34 14
	h x≷y	35 52
	g x≤y	32 71
	GTO 4	22 04
	RCL A	34 11
	GTO f b	22 31 12
	g LBLf e	32 25 15
110	.	83
	0	00
	1	01

STEP	KEY ENTRY	KEY CODE
	f %	31 82
	STO 3	33 03
	2	02
	÷	81
	-	51
	STO 1	33 01
	STO 0	33 00
120	f GSB 8	31 22 08
	STO 4	33 04
	RCL 1	34 01
	RCL 3	34 03
	+	61
	STO 0	33 00
	f GSB 8	31 22 08
	STO 2	33 02
	RCL 4	34 04
	-	51
130	RCL 3	34 03
	÷	81
	GTO f d	22 31 14
	g LBLf b	32 25 12
	f FIX	31 23
	GTO f e	22 31 15
	g LBLf d	32 25 14
	RCL 2	34 02
	GTO 2	22 02
	f LBL 5	31 25 05
140	RCL 0	34 00
	f GSB 8	31 22 08
	STO 2	33 02
	f LBL 2	31 25 02
	RCL 1	34 01
	RCL 0	34 00
	STO 1	33 01
	-	51
	RCL 4	34 04
	RCL 2	34 02
150	STO 4	33 04
	-	51
	÷	81
	x	71
	STO -0	33 51 00
	RCL 0	34 00
	h PAUSE	35 72
	÷	81
	f RND	31 24
	f x≠0	31 61
160	GTO 5	22 05
	RCL 0	34 00
	GTO 3	22 03
	h RTN	35 22
	f LBL 8	31 25 08
	RCL D	34 14
	1	01
	.	83
	4	04
	÷	81
170	RCL 0	34 00
	÷	81
	f LOG	31 53
	RCL C	34 13
	x	71
	4	04
	x	71
	RCL 0	34 00
	-	51
	h RTN	35 22
180	f LBL 4	31 25 04
	RCL A	34 11
	GTO 3	22 03
	h RTN	35 22
	f LBL A	31 25 11
	f P≷S	31 42
	RCL 1	34 01
	EEX	43
	CHS	42
	6	06
190	ENTER ↑	41
	h R↓	35 53
	x	71
	STO B	33 12
	RCL 2	34 02
	STO C	33 13
	f P≷S	31 42
	RCL 6	34 06
	f $\sqrt{x}$	31 54
	RCL 0	34 00
200	x	71
	.	83
	8	08
	7	07
	1	01
	÷	81
	RCL 2	34 02
	g x^2	32 54
	RCL 3	34 03
	g x^2	32 54
210	-	51
	f $\sqrt{x}$	31 54
	÷	81
	.	83
	3	03
	7	07
	5	05
	h y^x	35 63
	h RTN	35 22
	f LBL C	31 25 13
220	RCL D	34 14
	f -x-	31 84
	RCL E	34 15
	f -x-	31 84
	h RTN	35 22

STEP	KEY ENTRY	KEY CODE
001	f LBL A	31 25 11
	STO (i)	33 24
	f GSB 1	31 22 01
	h RTN	35 22
	f LBL 1	31 25 01
	f ISZ	31 34
	R/S	84
	h ST I	35 33
	GTO 1	22 01
010	h RTN	35 22
	f LBL B	31 25 12
	RCL 0	34 00
	RCL 4	34 04
	x	71
	.	83
	3	03
	2	02
	3	03
	x	71
020	RCL 2	34 02
	÷	81
	STO 5	33 05
	RCL 1	34 01
	.	83
	0	00
	1	01
	1	01
	9	09
	x	71
030	RCL 2	34 02
	g x^2	32 54
	÷	81
	f -x-	31 84
	STO A	33 11
	f GSB 9	31 22 09
	h PAUSE	35 72
	RCL A	34 11
	g x^2	32 54
	RCL D	34 14
040	x	71
	RCL 5	34 05
	x	71
	STO E	33 15
	GTO C	22 13
	f LBL 9	31 25 09
	f GSB 8	31 22 08
	RCL B	34 12
	.	83
	1	01
050	3	03
	5	05
	g x≤y	32 71
	GTO 2	22 02
	h x≷y	35 52
	h 1/x	35 62
	.	83

STEP	KEY ENTRY	KEY CODE
	0	00
	0	00
	2	02
060	0	00
	7	07
	x	71
	STO D	33 14
	1	01
	h RTN	35 22
	f LBL 2	31 25 02
	RCL B	34 12
	.	83
	4	04
070	g x≤y	32 71
	GTO 3	22 03
	h x≷y	35 52
	.	83
	3	03
	1	01
	9	09
	2	02
	CHS	42
	h y^x	35 63
080	.	83
	0	00
	0	00
	8	08
	1	01
	4	04
	x	71
	STO D	33 14
	2	02
	h RTN	35 22
090	f LBL 3	31 25 03
	RCL B	34 12
	RCL C	34 13
	g x≤y	32 71
	GTO 4	22 04
	h x≷y	35 52
	h 1/x	35 62
	.	83
	3	03
	5	05
100	5	05
	h y^x	35 63
	.	83
	0	00
	0	00
	6	06
	6	06
	2	02
	x	71
	.	83
110	0	00
	0	00
	1	01

TPF-D-007

STEP	KEY ENTRY	KEY CODE
	8	08
	+	61
	STO D	33 14
	3	03
	h RTN	35 22
	f LBL 4	31 25 04
	RCL 6	34 06
120	g x^2	32 54
	h 1/x	35 62
	STO D	33 14
	4	04
	h RTN	35 22
	f LBL 8	31 25 08
	.	83
	0	00
	1	01
	1	01
130	9	09
	RCL 1	34 01
	x	71
	RCL 2	34 02
	RCL 3	34 03
	x	71
	÷	81
	f -x-	31 84
	STO B	33 12
	RCL 2	34 02
140	3	03
	.	83
	7	07
	x	71
	.	83
	0	00
	0	00
	0	00
	7	07
	5	05
150	÷	81
	STO 7	33 07
	1	01
	.	83
	0	00
	5	05
	h y^x	35 63
	1	01
	.	83
	4	04
160	x	71
	RCL 7	34 07
	f LOG	31 53
	4	04
	x	71
	STO 6	33 06
	x	71
	7	07
	7	07

STEP	KEY ENTRY	KEY CODE
	4	04
170	2	02
	÷	81
	STO C	33 13
	f -x-	31 84
	h RTN	35 22
	g LBLf a	32 25 11
	f CL REG	31 43
	h RTN	35 22
	g LBLf b	32 25 12
	.	83
180	7	07
	÷	81
	2	02
	4	04
	x	71
	STO 1	33 01
	h PAUSE	35 72
	h RTN	35 22
	f LBL C	31 25 13
	RCL E	34 15
190	h RTN	35 22
	f LBL D	31 25 14
	RCL 1	34 01
	.	83
	0	00
	2	02
	9	09
	2	02
	x	71
	h PAUSE	35 72
200	RCL E	34 15
	x	71
	1	01
	7	07
	1	01
	5	05
	÷	81
	.	83
	7	07
	7	07
210	÷	81
	h RTN	35 22

STEP	KEY ENTRY	KEY CODE
001	f LBL A	31 25 11
	STO (i)	33 24
	f GSB 1	31 22 01
	h RTN	35 22
	f LBL 1	31 25 01
	f ISZ	31 34
	R/S	84
	h ST I	35 33
	GTO 1	22 01
010	h RTN	35 22
	f LBL B	31 25 12
	2	02
	5	05
	.	83
	7	07
	9	09
	RCL 0	34 00
	RCL 5	34 05
	x	71
020	x	71
	RCL 6	34 06
	4	04
	6	06
	0	00
	+	61
	÷	81
	RCL A	34 11
	2	02
	.	83
030	5	05
	h y^x	35 63
	÷	81
	g x^2	32 54
	RCL 2	34 02
	f P≷S	31 42
	RCL 0	34 00
	x	71
	x	71
	RCL 1	34 01
040	4	04
	6	06
	0	00
	+	61
	x	71
	STO 8	33 08
	f P≷S	31 42
	.	83
	0	00
	3	03
050	7	07
	5	05
	RCL 9	34 09
	RCL 2	34 02
	x	71
	x	71
	f P≷S	31 42
	RCL 1	34 01
	4	04
	6	06
060	0	00
	+	61
	÷	81
	STO 9	33 09
	f P≷S	31 42
	h RTN	35 22
	f LBL C	31 25 13
	RCL 0	34 00
	f P≷S	31 42
	STO 4	33 04
070	f P≷S	31 42
	RCL 1	34 01
	÷	81
	RCL 2	34 02
	h ST I	35 33
	x	71
	RCL 3	34 03
	RCL 4	34 04
	2	02
	x	71
080	-	51
	STO A	33 11
	÷	81
	RCL 5	34 04
	x	71
	RCL 6	34 06
	4	04
	6	06
	0	00
	+	61
090	÷	81
	.	83
	4	04
	7	07
	7	07
	5	05
	x	71
	f -x-	31 84
	STO B	33 12
	f GSB B	31 22 12
100	RCL A	34 11
	RCL 7	34 07
	EEX	43
	CHS	42
	6	06
	ENTER ↑	41
	h R↓	35 53
	x	71
	÷	81
	3	03
110	.	83
	7	07
	x	71

STEP	KEY ENTRY	KEY CODE
	STO D	33 14
	f LOG	31 53
	4	04
	x	71
	STO C	33 13
	f -x-	31 84
	RCL D	34 14
120	RCL 8	34 08
	h 1/x	35 62
	h y^x	35 63
	RCL C	34 13
	x	71
	1	01
	.	83
	4	04
	x	71
	STO D	33 14
130	f -x-	31 84
	RCL B	34 12
	h x≷y	35 52
	g x≤y	32 71
	GTO 4	22 04
	RCL C	34 13
	GTO f b	22 31 12
	g LBLf e	32 25 15
	.	83
	0	00
140	1	01
	f %	31 82
	STO C	33 03
	2	02
	÷	81
	-	51
	STO 1	33 01
	STO 0	33 00
	f GSB 8	31 22 08
	STO 4	33 04
150	RCL 1	34 01
	RCL 3	34 03
	+	61
	STO 0	33 00
	f GSB 8	31 22 08
	STO 2	33 02
	RCL 4	34 04
	-	51
	RCL 3	34 03
	÷	81
160	GTO f d	22 31 14
	g LBLf b	32 25 12
	f FIX	31 23
	GTO f e	22 31 15
	g LBLf d	32 25 14
	RCL 2	34 02
	GTO 2	22 02
	f LBL 5	31 25 05
	RCL 0	34 00
	f GSB 8	31 22 08
170	STO 2	33 02
	f LBL 2	31 25 02
	RCL 1	34 01
	RCL 0	34 00
	STO 1	33 01
	-	51
	RCL 4	34 04
	RCL 2	34 02
	STO 4	33 04
	-	51
180	÷	81
	x	71
	STO -0	33 51 00
	RCL 0	34 00
	h PAUSE	35 72
	÷	81
	f RND	31 24
	f x≠0	31 61
	GTO 5	22 05
	RCL 0	34 00
190	h 1/x	35 62
	g x^2	32 54
	f P≷S	31 42
	RCL 8	34 08
	x	71
	STO 8	33 08
	f P≷S	31 42
	h RTN	35 22
	f LBL 8	31 25 08
	RCL B	34 12
200	1	01
	.	83
	4	04
	÷	81
	RCL 0	34 00
	÷	81
	f LOG	31 53
	RCL 8	34 08
	x	71
	4	04
210	x	71
	RCL 0	34 00
	-	51
	h RTN	35 22
	f LBL 4	31 25 04
	RCL C	34 13
	h 1/x	35 62
	g x^2	32 54
	f P≷S	31 42
	RCL 8	34 08
220	x	71
	STO 8	33 08
	f P≷S	31 42
	h RTN	35 22

STEP	KEY ENTRY	KEY CODE
001	f LBL A	31 25 11
	f P≷S	31 42
	RCL 4	34 04
	RCL 0	34 00
	f $\sqrt{x}$	31 54
	x	71
	.	83
	8	08
	7	07
010	1	01
	÷	81
	RCL 5	34 05
	2	02
	.	83
	6	06
	7	07
	h y^x	35 63
	÷	81
	g x^2	32 54
020	CHS	42
	RCL 2	34 02
	g x^2	32 54
	+	61
	f $\sqrt{x}$	31 54
	f -x-	31 84
	STO 5	33 05
	f P≷S	31 42
	h RTN	35 22
	f LBL C	31 25 13
030	h RC I	35 34
	4	04
	3	03
	x	71
	CHS	42
	6	06
	9	09
	6	06
	.	83
	5	05
040	+	61
	STO A	33 11
	h RC I	35 34
	3	03
	0	00
	0	00
	x	71
	1	01
	8	08
	0	00
050	+	61
	STO B	33 12
	f P≷S	31 42
	RCL 1	34 01
	f P≷S	31 42
	4	04
	6	06

STEP	KEY ENTRY	KEY CODE
	0	00
	+	61
	RCL B	34 12
060	÷	81
	STO C	33 13
	h RTN	35 22
	f LBL D	31 25 14
	f P≷S	31 42
	RCL 2	34 02
	RCL 3	34 03
	+	61
	3	03
	h y^x	35 63
070	f P≷S	31 42
	RCL 0	34 00
	f P≷S	31 42
	RCL 3	34 03
	+	61
	3	03
	h y^x	35 63
	-	51
	STO 7	33 07
	RCL 2	34 02
080	RCL 3	34 03
	+	61
	g x^2	32 54
	RCL 3	34 03
	f P≷S	31 42
	RCL 0	34 00
	+	61
	g x^2	32 54
	-	51
	f P≷S	31 42
090	h x≷y	35 52
	÷	81
	.	83
	6	06
	7	07
	x	71
	RCL 3	34 03
	-	51
	f P≷S	31 42
100	STO D	33 14
	h RTN	35 22
	f LBL E	31 25 15
	RCL D	34 14
	RCL A	34 11
	h PAUSE	35 72
	÷	81
	STO E	33 15
	RCL C	34 13
	.	83
110	3	03
	4	04
	1	01

STEP	KEY ENTRY	KEY CODE
	x	71
	.	83
	6	06
	0	00
	4	04
	CHS	42
	+	61
120	RCL E	34 15
	x	71
	1	01
	+	61
	STO E	33 15
	h RTN	35 22
	g LBLf b	32 25 12
	f P≷S	31 42
	RCL 8	34 08
	RCL E	34 15
130	x	71
	f P≷S	31 42
	RCL 0	34 00
	f P≷S	31 42
	RCL 3	34 03
	+	61
	g x^2	32 54
	+	61
	STO 6	33 06
	RCL 9	34 09
140	RCL D	34 14
	g x^2	32 54
	x	71
	RCL E	34 15
	÷	81
	RCL 6	34 06
	+	61
	f $\sqrt{x}$	31 54
	CHS	42
	RCL 2	34 02
150	RCL 3	34 03
	+	61
	+	61
	f P≷S	31 42
	h RTN	35 22
	f LBL B	31 25 12
	RCL A	34 11
	f P≷S	31 42
	STO 5	33 05
	f P≷S	31 42
160	f GSB C	31 22 13
	f GSB A	31 22 11
	GTO f b	22 31 12
	g LBLf e	32 25 15
	.	83
	0	00
	1	01
	f %	31 82
	STO 3	33 03

STEP	KEY ENTRY	KEY CODE
	2	02
170	÷	81
	-	51
	STO 1	33 01
	STO 0	33 00
	f GSB 8	31 22 08
	STO 4	33 04
	RCL 1	34 01
	RCL 3	34 03
	+	61
	STO 0	33 00
180	f GSB 8	31 22 08
	STO 2	33 02
	RCL 4	34 04
	-	51
	RCL 3	34 03
	÷	81
	GTO f d	22 31 14
	g LBLf b	32 25 12
	f FIX	31 23
	GTO f e	22 31 15
190	g LBLf d	32 25 14
	RCL 2	34 02
	GTO 2	22 02
	f LBL 5	31 25 05
	RCL 0	34 00
	f GSB 8	31 22 08
	STO 2	33 02
	f LBL 2	31 25 02
	RCL 1	34 01
	RCL 0	34 00
200	STO 1	33 01
	-	51
	RCL 4	34 04
	RCL 2	34 02
	STO 4	33 04
	-	51
	÷	81
	x	71
	STO -0	33 51 00
	RCL 0	34 00
210	h PAUSE	35 72
	÷	81
	f RND	31 24
	f x≠0	31 61
	GTO 5	22 05
	RCL 0	34 00
	h RTN	35 22
	f LBL 8	31 25 08
	f GSB D	31 22 14
	f GSB E	31 22 15
220	g GSB b	32 22 12
	h RTN	35 22

TPF-D-009

STEP	KEY ENTRY	KEY CODE
001	f LBL A	31 25 11
	STO (i)	33 24
	f GSB 1	31 22 01
	h RTN	35 22
	f LBL 1	31 25 01
	f ISZ	31 34
	R/S	84
	h ST I	35 33
	GTO 1	22 01
010	h RTN	35 22
	f LBL B	31 25 12
	RCL 0	34 00
	RCL 1	34 01
	÷	81
	f $\sqrt{x}$	31 54
	STO A	33 11
	f -x-	31 84
	.	83
	4	04
020	9	09
	h y^x	35 63
	2	02
	4	04
	.	83
	4	04
	8	08
	x	71
	STO B	33 12
	f -x-	31 84
030	RCL 2	34 02
	4	04
	2	02
	x	71
	RCL 3	34 03
	8	08
	.	83
	3	03
	3	03
	x	71
040	x	71
	2	02
	4	04
	÷	81
	f GSB 2	31 22 02
	2	02
	4	04
	÷	81
	g x^2	32 54
	h π	35 73
050	x	71
	÷	81
	f -x-	31 84
	STO C	33 13
	RCL 6	34 06
	1	01
	0	00

STEP	KEY ENTRY	KEY CODE
	0	00
	0	00
	x	71
060	RCL 7	34 07
	2	02
	9	09
	x	71
	x	71
	2	02
	4	04
	÷	81
	3	03
	7	07
070	9	09
	÷	81
	f P≷S	31 42
	RCL 5	34 05
	f P≷S	31 42
	2	02
	4	04
	÷	81
	g x^2	32 54
	h π	35 73
080	x	71
	÷	81
	STO D	33 14
	f -x-	31 84
	RCL 8	34 08
	.	83
	0	00
	7	07
	5	05
	÷	81
090	RCL 9	34 09
	6	06
	2	02
	.	83
	3	03
	÷	81
	x	71
	f $\sqrt{x}$	31 54
	STO E	33 15
	f -x-	31 84
100	f P≷S	31 42
	RCL 0	34 00
	f P≷S	31 42
	4	04
	8	08
	.	83
	6	06
	4	04
	x	71
	1	01
110	2	02
	CHS	42
	+	61

STEP	KEY ENTRY	KEY CODE
	f GSB 3	31 22 03
	h R↑	35 54
	f -x-	31 84
	x	71
	f P≷S	31 42
	STO 3	33 03
	f P≷S	31 42
120	7	07
	3	03
	h x≷y	35 52
	÷	81
	h ST I	35 33
	6	06
	2	02
	.	83
	3	03
	RCL 9	34 09
130	÷	81
	g x^2	32 54
	f P≷S	31 42
	RCL 2	34 02
	f P≷S	31 42
	x	71
	.	83
	3	03
	4	04
	h y^x	35 63
140	h RC I	35 34
	x	71
	f P≷S	31 42
	STO 4	33 04
	f P≷S	31 42
	f -x-	31 84
	RCL D	34 14
	RCL E	34 15
	÷	81
	STO A	33 11
150	f -x-	31 84
	h 1/x	35 62
	f P≷S	31 42
	RCL 4	34 04
	f P≷S	31 42
	RCL C	34 13
	x	71
	x	71
	STO B	33 12
	f -x-	31 84
160	RCL D	34 14
	3	03
	6	06
	0	00
	0	00
	÷	81
	RCL 8	34 08
	÷	81
	f P≷S	31 42

STEP	KEY ENTRY	KEY CODE
	STO 6	33 06
170	f P≷S	31 42
	f -x-	31 84
	RCL C	34 13
	3	03
	6	06
	0	00
	0	00
	÷	81
	RCL 9	34 09
	÷	81
180	f P≷S	31 42
	STO 7	33 07
	f P≷S	31 42
	f -x-	31 84
	h RTN	35 22
	f LBL 2	31 25 02
	RCL 4	34 04
	RCL 5	34 05
	2	02
	x	71
190	-	51
	f P≷S	31 42
	STO 5	33 05
	f P≷S	31 42
	h RTN	35 22
	f LBL 3	31 25 03
	7	07
	.	83
	4	04
	CHS	42
200	EEX	43
	CHS	42
	4	04
	ENTER ↑	41
	f P≷S	31 42
	RCL 1	34 01
	f P≷S	31 42
	x	71
	g e^x	32 52
	9	09
210	9	09
	.	83
	4	04
	6	06
	x	71
	1	01
	0	00
	0	00
	÷	81
	f -x-	31 84
220	h RTN	35 22
	f LBL C	31 25 13
	f CL REG	31 43
	h RTN	35 22

TPF-D-010

STEP	KEY ENTRY	KEY CODE
001	f LBL A	31 25 11
	h ST I	35 33
	f GSB (i)	31 22 24
	h RTN	35 22
	f LBL 1	31 25 01
	f GSB 9	31 22 09
	.	83
	8	08
	1	01
010	5	05
	h y^x	35 63
	1	01
	1	01
	9	09
	0	00
	x	71
	RCL C	34 13
	f $\sqrt{x}$	31 54
	÷	81
020	f GSB 8	31 22 08
	h RTN	35 22
	f LBL 9	31 25 09
	RCL 0	34 00
	RCL 1	34 01
	÷	81
	f $\sqrt{x}$	31 54
	h RTN	35 22
	f LBL 8	31 25 08
	g x^2	32 54
030	RCL 1	34 01
	x	71
	h RTN	35 22
	f LBL 2	31 25 02
	f GSB 9	31 22 09
	1	01
	5	05
	4	04
	0	00
	0	00
040	x	71
	RCL C	34 13
	.	83
	8	08
	h y^x	35 63
	÷	81
	f GSB 8	31 22 08
	h RTN	35 22
	f LBL 3	31 25 03
	f GSB 9	31 22 09
050	.	83
	8	08
	5	05
	5	05
	h y^x	35 63
	2	02
	7	07

STEP	KEY ENTRY	KEY CODE
	.	83
	3	03
	1	01
060	3	03
	x	71
	RCL C	34 13
	.	83
	1	01
	7	07
	h y^x	35 63
	÷	81
	f GSB 8	31 22 08
	h RTN	35 22
070	f LBL 4	31 25 04
	f GSB 9	31 22 09
	.	83
	7	07
	5	05
	h y^x	35 63
	1	01
	4	04
	.	83
	2	02
080	x	71
	RCL C	34 13
	.	83
	1	01
	h y^x	35 63
	÷	81
	f GSB 8	31 22 08
	h RTN	35 22
	f LBL 5	31 25 05
	f P≷S	31 42
090	RCL 5	34 05
	f P≷S	31 42
	1	01
	0	00
	h x≷y	35 52
	g x>y	32 81
	GTO 0	22 00
	g LBLf a	32 25 11
	h PAUSE	35 72
	.	83
100	0	00
	2	02
	1	01
	x	71
	CHS	42
	.	83
	3	03
	4	04
	3	03
	+	61
110	f GSB 9	31 22 09
	h x≷y	35 52
	h y^x	35 63

STEP	KEY ENTRY	KEY CODE
	f P≷S	31 42
	RCL 5	34 05
	f P≷S	31 42
	.	83
	3	03
	1	01
	2	02
120	5	05
	x	71
	CHS	42
	4	04
	.	83
	8	08
	+	61
	x	71
	f GSB 8	31 22 08
	h RTN	35 22
130	f LBL 0	31 25 00
	1	01
	0	00
	f P≷S	31 42
	STO 5	33 05
	f P≷S	31 42
	GTO f a	22 31 11
	h RTN	35 22
	f LBL 6	31 25 06
	RCL D	34 14
140	3	03
	6	06
	0	00
	0	00
	÷	81
	g x^2	32 54
	f P≷S	31 42
	RCL 8	34 08
	f P≷S	31 42
	x	71
150	1	01
	9	09
	3	03
	.	83
	2	02
	÷	81
	RCL 8	34 08
	÷	81
	f P≷S	31 42
	RCL 9	34 09
160	f P≷S	31 42
	1	01
	2	02
	÷	81
	÷	81
	f GSB 7	31 22 07
	x	71
	h RTN	35 22
	f LBL 7	31 25 07

STEP	KEY ENTRY	KEY CODE
	RCL C	34 13
170	RCL D	34 14
	÷	81
	f P≷S	31 42
	RCL 2	34 02
	x	71
	RCL 3	34 03
	÷	81
	f P≷S	31 42
	.	83
	2	02
180	h y^x	35 63
	.	83
	0	00
	0	00
	4	04
	3	03
	5	05
	x	71
	h RTN	35 22
	f LBL E	31 25 15
190	f P≷S	31 42
	STO 9	33 09
	R/S	84
	STO 8	33 08
	R/S	84
	STO 7	33 07
	R/S	84
	STO 4	33 04
	R/S	84
	STO 3	33 03
200	f P≷S	31 42
	h RTN	35 22
	f LBL B	31 25 12
	RCL 9	34 09
	f P≷S	31 42
	RCL 7	34 07
	RCL 4	34 04
	x	71
	x	71
	1	01
210	4	04
	4	04
	÷	81
	1	01
	.	83
	6	06
	RCL 6	34 06
	.	83
	7	07
	CHS	42
220	h y^x	35 63
	x	71
	x	71
	f P≷S	31 42
	h RTN	35 22

3.3 Martinelli Correlation Computation

Introduction

This program evaluates two-phase flow pressure drop and percent hold-up in a horizontal system. Reynolds numbers for both fluids are in the turbulent region, ie >2000.

Theoretical Formulae

$$\Delta P_{TP} = \Delta P_G \, \phi g^2$$

$$\begin{aligned}\phi g = \exp\ & (.168101 \times 10^{-4}) \ln^6 x \\ & + (.305664 \times 10^{-4}) \ln^5 x \\ & - (.949663 \times 10^{-3}) \ln^4 x \\ & - (.113212 \times 10^{-2}) \ln^3 x \\ & + (.615325 \times 10^{-1}) \ln^2 x \\ & + (.495481) \ln x \\ & + 1.44105\end{aligned}$$

$$\text{Percent hold-up} = y = 27.35\, x^{.72} \quad (0 \leq x \leq .8)$$
$$y = 26.27\, x^{.32} \quad (x > .8)$$

$$x = \sqrt{\frac{\Delta P_L}{\Delta P_G}}$$

TPF-D-011 - Martinelli Correlation Computation

Input:
(1) ΔP for liquid - psi (over total length)
(2) ΔP for gas - psi (over total length)

Computes:
(1) ΔP_{TP} - psi
(2) % hold-up - n

Note: This correlation is valid for line sizes "1" thru "6". Also, no elevation pressure drop is computed.

STEP	INSTRUCTIONS	KEYS	OUTPUT UNITS
1.	load card TPF-D-011, side 1 & 2		
2.	input pressure drop due to liquid	fa	psi
3.	input pressure drop due to gas	fb	psi
4.	compute horizontal pressure drop	A	psi
5.	compute percent liquid hold-up	B	n
6.	reset for new case	f CLR Reg	
7.	Option	RCL C	$\sqrt{\frac{\Delta P_L}{\Delta P_G}}$

Example: **TPF-D-011** - Martinelli Correlation

Given Conditions:

Assume that P_L = .23 psi

P_G = 10.8 psi

Execution: Load card TPF-D-011, side 1 & 2

Store parameters in this order:

KEYSTROKES	OUTPUT
.23 "fa"	.23
10.8 "fb"	10.8

To Compute:

Press "A"	44.72 psi
Press "B"	6.84%

STEP	KEY ENTRY	KEY CODE
001	f LBL A	31 25 11
	RCL 7	34 07
	RCL 8	34 08
	STO D	33 14
	÷	81
	f $\sqrt{x}$	31 54
	STO C	33 13
	f LN	31 52
	STO E	33 15
010	0	00
	STO 9	33 09
	f GSB 1	31 22 01
	f GSB 2	31 22 02
	f GSB 3	31 22 03
	f GSB 4	31 22 04
	f GSB 5	31 22 05
	f GSB 6	31 22 06
	RCL 9	34 09
	1	01
020	.	83
	4	04
	4	04
	1	01
	0	00
	5	05
	+	61
	g e^x	32 52
	g x^2	32 54
	RCL D	34 14
030	x	71
	h RTN	35 22
	f LBL 1	31 25 01
	.	83
	1	01
	6	06
	8	08
	1	01
	0	00
	1	01
040	EEX	43
	4	04
	CHS	42
	ENTER↑	41
	STO A	33 11
	RCL E	34 15
	6	06
	h y^x	35 63
	RCL A	34 11
	x	71
050	STO 9	33 09
	h RTN	35 22
	f LBL 2	31 25 02
	.	83
	3	03
	0	00
	5	05
	6	06
	6	06
	4	04
060	EEX	43
	4	04
	CHS	42
	ENTER↑	41
	STO A	33 11
	RCL E	34 15
	5	05
	h y^x	35 63
	RCL A	34 11
	x	71
070	STO +9	33 61 09
	h RTN	35 22
	f LBL 3	31 25 03
	.	83
	9	09
	4	04
	9	09
	6	06
	3	03
	EEX	43
080	3	03
	CHS	42
	ENTER↑	41
	CHS	42
	STO A	33 11
	RCL E	34 15
	4	04
	h y^x	35 63
	RCL A	34 11
	x	71
090	STO +9	33 61 09
	h RTN	35 22
	f LBL 4	31 25 04
	.	83
	1	01
	1	01
	3	03
	2	02
	1	01
	2	02
100	EEX	43
	CHS	42
	2	02
	ENTER↑	41
	CHS	42
	STO A	33 11
	RCL E	34 15
	3	03
	h y^x	35 63
	RCL A	34 11
110	x	71
	STO +9	33 61 09
	h RTN	35 22

STEP	KEY ENTRY	KEY CODE
	f LBL 5	31 25 05
	.	83
	6	06
	1	01
	5	05
	3	03
	2	02
120	5	05
	EEX	43
	1	01
	CHS	42
	ENTER↑	41
	STO A	33 11
	RCL E	34 15
	g x^2	32 54
	RCL A	34 11
	x	71
130	STO +9	33 61 09
	h RTN	35 22
	f LBL 6	31 25 06
	.	83
	4	04
	9	09
	5	05
	4	04
	8	08
	1	01
140	ENTER↑	41
	RCL E	34 15
	x	71
	STO +9	33 61 09
	h RTN	35 22
	f LBL B	31 25 12
	RCL C	34 13
	.	83
	8	08
	g x≤y	32 71
150	f LBL 8	31 25 08
	RCL C	34 13
	.	83
	7	07
	2	02
	h y^x	35 63
	2	02
	7	07
	.	83
	3	03
160	5	05
	x	71
	h RTN	35 22
	f LBL 8	31 25 08
	RCL C	34 13
	.	83
	3	03
	2	02
	h y^x	35 63
	2	02
170	6	06
	.	83
	2	02
	7	07
	x	71
	h RTN	35 22
	g LBLf a	32 25 11
	STO 7	33 07
	h RTN	35 22
	g LBLf b	32 25 12
180	STO 8	33 08
	h RTN	35 22

References

3-1 Ovid Baker, "Designing Pipelines for Simultaneous Flow of Oil and Gas", Pipeline Engineer, February, 1960.

3-2 John R. Campbell, "Problems of Multiphase Pipeline Flow", Flow Calculations Handbook, Oil & Gas Journal, 1960.

3-3 Dukler, A.E., et al, "Frictional Pressure Drop in Two-Phase Flow: A comparison of Existing Correlations for Pressure Loss and Holdup". An Approach Through Similarity Analysis A.I.C.H.E. Journal, January 1965.

3-4 Don L. Love, "Improve Two-Phase Flow Calculations", Oil & Gas Journal, January 7, 1974.

3-5 A.E. DeGame et al., "Pressure Drop Sample Calculations", Chemical Engineering, November 2, 1970.

3-6 K. Aziz et al., "Flow of Complex Mixtures in Pipes", Reinhold Publishing Co., New York, 1972.

3-7 Gould, T.L. and Ramsey, E.L., "Design of Offshore Gas Pipelines Accounting for Two-Phase Flow", Journal of Petroleum Technology, March 1975.

3-8 Hope, P.M. and Nelson, R.G., "Improved Calculations Aid Design of Wet Gas Pipelines", The Oil & Gas Journal, Oct. 1977.

Nomenclature

API - oil gravity
A_f - cross sectional flow area - $ft.^2$
A - saturated oil viscosity coefficient
B - saturated oil viscosity coefficient
B_o - oil formation factor - BBL/STB
D - inside diameter - inches
d - inside pipe diameter - feet
f - friction factor
f_{tp} - friction factor for two-phase flow mixture
F_e - elevation factor
G - mass velocity of the gas - $lb./HR/ft.^2$
H - average height of hills - feet
k - a constant - 1.13
L - mass velocity of the liquid - $lb./HR/ft.^2$
L' - length of the pipeline - feet
LFV - liquid line fill - BBLS/mile
M - s.g. gas (2g) (s.g.) = MW
P_m - average line pressure - psia
ΔP_L - pressure drop for the liquid phase flowing in the pipe. The full cross-sectional area of the pipe is used in evaluating the pressure drop - PSI.
ΔP_G - pressure drop for the gas phase flowing in the pipe. The full cross-sectional area of the pipe is used in evaluating the pressure drop - PSI.
ΔP_{tp} - pressure drop for the two phases flowing simultaneously - PSI
Q_b - gas flow rate - MCFD
Rs - solution gas-oil ratio - $ft.^3/BBL$
T - absolute temperature of flowing gas - °R = °F + 460
U_1 - percent of pipe volume occupied by liquid at gas flowrate one
U_2 - percent of pipe volume occupied by liquid at gas flowrate two
V_g - superficial gas velocity in pipeline, gas volume at pipeline temperature and pressure, area for flow, the entire pipe cross section.
x - $(\Delta P_L \text{ (PSI)}/\Delta P_G \text{ (PSI)})^{.5}$
z - compressibility factor of flowing gas
ΦGTT - $(\Delta P_{TP}/\Delta P_G)^{.5}$ where both liquid and gas phases are in turbulent flow.
ρ_G - density of gas phase - $lb./ft.^3$
ρ_L - density of liquid phase - $lb./ft.^3$
λ - $((\rho_G/0.075)(\rho_L/62.3))^{1/2}$
ψ - $(73/\gamma)(\mu_L(62.3/\rho_L)^2)^{1/3}$
γ - surface tension of liquid phase, dynes per centimeter
μ_L - viscosity of liquid phase, centipoise
μ_{od} - viscosity of dead oil - centipoise
μ_{os} - viscosity of saturated oil - centipoise
μ_G - viscosity of gas - lb.m./ft.-sec.
μ_g - viscosity of gas - centipoise
V_1 - superficial liquid velocity in pipeline, liquid volume at pipeline temperature and pressure, area for flow, the entire pipe cross section.

4.0 Slurry Pipeline Design

4.1 Introduction

There are eight basic slurry design programs. They are divided into these principal parts:

SP-D-001 SP-D-002	Slurry Characteristics
SP-D-003	Settling Velocity Drag Coefficient
SP-D-004	C/C_A Characteristics, ie. determination of homogeneous and/or heterogeneous flow
SP-D-005	Pressure drop for homogeneous flow
SP-D-006A SP-D-006B SP-D-007	Two-Phase Vehicle Pressure Drop

4.2 Theoretical Formulae

A. Settling Velocity and Drag Coefficients

(1) Stokes Law: $Re<1$

$$\omega = \frac{g}{18} \frac{(\rho_s - \rho_f)}{\rho_f \nu} d^2$$

$$Re = \frac{\omega d}{\nu} \quad ; \quad C_D = \frac{24}{Re}$$

(2) Intermediate Law:

$$\frac{C_D}{Re} = \frac{4g (\rho_s - \rho_f) \nu}{3 \rho_f \omega^3}$$

$$C_D Re^2 = \frac{4g (\rho_s - \rho_f) d^3}{3 \rho_f \nu^2}$$

(3) Newton's Law: $1000 < Re < 2 \times 10^5$

$$\omega = \left[\frac{3.33 (\rho_s - \rho_f) d}{\rho_f} \right]^{.5} \quad ; \quad R = \frac{\omega d}{\nu} , \; C_D = .4$$

B. Characteristics

$$m = \frac{100}{\frac{C_w}{\rho_s} + \frac{100 - C_w}{\rho_f}}$$

$$C_v = \frac{C_w \, \rho_m}{\rho_s}$$

$$C_w = \frac{C_v \, \rho_s}{\rho_m} = 100 \, \frac{\rho_s}{\rho_m} \frac{(\rho_m - \rho_f)}{(\rho_s - \rho_f)}$$

$$\Phi = C_v/100$$

$$\frac{\mu_m}{\mu_o} = 1.25\Phi + 10.05\Phi^2 + (.00273) \exp(16.6 \, \Phi)$$

$$(C_p)m = \frac{(C_p)s \, (C_w)s + (C_p)f \, (C_w)f}{100}$$

$$k_m = k_1 \left[\frac{2k_f + k_s - 2\Phi \, (k_f - k_s)}{2k_f + k_s + \Phi \, (K_1 - k_s)} \right]$$

Volume of slurry = $\frac{\text{volume of solids}}{\text{concentration by volume}}$

Volume of water = volume of slurry - volume of solids

Density of slurry = $\frac{\text{weight of water \& solids}}{\text{volume of slurry}}$

C. C/C_A Ratio

$$\log_{10} C/C_A = \frac{-(1.8 \, w)}{(\beta \xi U^*)}$$

$$U^* = v \sqrt{f/2}$$

D. Homogeneous Flow $C/C_A > .8$ (Bingham Plastic Model)

Hedstrom number = $\frac{D^2 \, \tau_o \, \rho}{\eta^2}$

$$V_T = \frac{(\text{Re critical}) \, \eta}{D \, \rho}$$

$$\Delta PSI/mi. = \frac{.241\ f\ \rho_m\ Q^2}{D^5}$$

E. Two-Phase Vehicle Parameters

$$F_L = \frac{V_D}{\sqrt{2gD\ (s - 1)}}$$

$$V_D = F_L \left[2gD\ \frac{(\rho_s - \rho_f)}{\rho_f} \right]^{.5}$$

$$V_D = F_L \left[2gD\ \frac{(\rho_s - \rho_f)}{\rho_f} \right]^{.5} (d/D)^{1/6}$$

$$f_m = f_f \left[1 + k' \left[\frac{(gD)}{v^2}\ \frac{(\rho_s - \rho_f)}{\rho_f}\ \frac{1}{\sqrt{C_D}} \right]^{3/2} \Phi \right]$$

$$\Delta p\ \text{bed} = 82 \times \Delta p\ \text{water} \times \Phi\ \text{bed} \left[\frac{gD\ (s - 1)}{v^2\ \sqrt{C_D}} \right]^{1.5}$$

C/C_A - Heterogeneous flow $\lesssim .10$
Homogeneous flow $\gtrsim .8$

SP-D-001 - Slurry Characteristics

Input:

(1) solid density - n (ratio)
(2) liquid density - n (ratio)
(3) volume of solid to be transported - TONS/yr.
(4) percent solids by volume - %
(5) efficiency of pipeline - n≥1
(where 1 = 100%
1.1 = 90%)

Computes:

(1) flowrate of solids - $ft.^3$/minute
(2) flowrate of slurry - $ft.^3$/minute
(3) flowrate of water - $ft.^3$/minute
(4) flowrate of water - lbs./minute
(5) density of mixture "slurry" - lbs./$ft.^3$
(6) density of mixture "slurry" - n (ratio)
(7) concentration of solids by weight - %
(8) Optional:
 (a) compute density of slurry - n (ratio)
 (b) compute all parameters with new vol.% - %
 (c) compute all parameters with new volume of solids - TONS/yr.
(9) An entire new case - Press "fa" to reset program

STEP	INSTRUCTIONS	KEYS	OUTPUT UNITS
1.	load card SP-D-001, side 1		
2.	input density of solids	A	n (ratio)
3.	input density of liquid	A	n (ratio)
4.	input volume of solids to be transported	A	TONS/yr.
5.	input solid volume precent	A	n%
6.	input efficiency of pipeline, n≥1	A	n
7.	Compute:	B	
	(1) flowrate of solids		$ft.^3$/minute
	(2) flowrate of slurry		$ft.^3$/minute
	(3) flowrate of water		$ft.^3$/minute
	(4) flowrate of water		lbs./minute
	(5) density of slurry		lbs./$ft.^3$
	(6) density of slurry		n (ratio)
	(7) concentration of solids in percent by weight		n%
8.	Optional:		
	(a) compute density of slurry only.	C	n (ratio)
	(b) compute all parameters with a new solid % by volume.	D B (go to step 7)	n%

STEP	INSTRUCTIONS	KEYS	OUTPUT UNITS
	(c) compute all parameters with a new solids volume flowrate.	E B (go to step 7)	TONS/yr.
9.	Do you wish to run a new case?	fa	(resets program)

Example: SP-D-001

Given Conditions:

solids to be transported: (coal) s.g. = 1.4
liquid to be transported: (water) s.g. = 1.0
volume of solid to be transported: 500,000 TONS/yr.
volume percent of solids: 40%
pipeline efficiency (90%): 1.1

Execution: Load card SP-D-001, side 1 only

Store parameters in this order:

KEYSTROKES	OUTPUT
1.4 "A"	1.40
1.0 "A"	1.00
500,000 "A"	500,000.00
40 "A"	40.00
1.1 "A"	1.10

To Compute:

Press "B"	"24.25" ft.^{3}/min. (flowrate of solids)
	"60.63" ft.^{3}/min. (flowrate of slurry)
	"36.38" ft.^{3}/min. (flowrate of water)
	"2,266.39" lbs./min. (flowrate of water)
	"72.268" lbs./ft.^{3} (density of slurry)
	"1.16" ratio (density of slurry)
	"48.28" % (solids concentration by weight)

Optional:

Press "C"	1.16 ratio (slurry density only)

To Compute all parameters with new volume solid percentage:

45 "D"	45.00
Press "B"	"24.25" ft.^{3}/min.
	"53.89" ft.^{3}/min.
	"29.64" ft.^{3}/min.
	"1,846.69" lbs./min.
	"73.51" lbs./ft.^{3}
	"1.18" ratio
	"53.39" %

To Compute all parameters with new solids volume flowrate:

KEYSTROKES	OUTPUT
1,300,000 "E"	1,300,000
40 "D"	40.00
Press "B"	"63.056" ft.^{3}/min.
	"157.64" ft.^{3}/min.
	"94.58" ft.^{3}/min.
	"5,892.62" lbs./min.
	"72.268" lbs./ft.^{3}
	"1.16" ratio
	"48.276" %

New case desired: to reset program

"fa" ________________ to reset program

SP-D-001

STEP	KEY ENTRY	KEY CODE
001	f LBL A	31 25 11
	STO (i)	33 24
	f GSB 1	31 22 01
	h RTN	35 22
	f LBL 1	31 25 01
	f ISZ	31 34
	R/S	84
	h ST I	35 33
	GTO 1	22 01
010	h RTN	35 22
	f LBL B	31 25 12
	RCL 2	34 02
	RCL 4	34 04
	x	71
	3	03
	.	83
	8	08
	4	04
	6	06
020	EEX	43
	CHS	42
	3	03
	ENTER ↑	41
	h R↓	35 53
	x	71
	RCL 0	34 00
	6	06
	2	02
	.	83
030	3	03
	x	71
	÷	81
	STO 5	33 05
	h PAUSE	35 72
	RCL 3	34 03
	1	01
	0	00
	0	00
	÷	81
040	÷	81
	STO 6	33 06
	h PAUSE	35 72
	RCL 5	34 05
	CHS	42
	+	61
	STO 7	33 07
	h PAUSE	35 72
	6	06
	2	02
050	.	83
	3	03
	x	71
	STO 9	33 09
	h PAUSE	35 72
	RCL 5	34 05
	RCL 0	34 00
	6	06
	2	02
	.	83
060	3	03
	x	71
	x	71
	+	61
	RCL 6	34 06
	÷	81
	h PAUSE	35 72
	6	06
	2	02
	.	83
070	3	03
	÷	81
	STO 8	33 08
	h PAUSE	35 72
	RCL 5	34 05
	RCL 0	34 00
	x	71
	6	06
	2	02
	.	83
080	3	03
	x	71
	STO +9	33 61 09
	RCL 9	34 09
	÷	81
	1	01
	0	00
	0	00
	x	71
	STO 9	33 09
090	h RTN	35 22
	f LBL C	31 25 13
	RCL 0	34 00
	1	01
	-	51
	RCL 3	34 03
	1	01
	0	00
	0	00
	÷	81
100	x	71
	1	01
	+	61
	h RTN	35 22
	f LBL D	31 25 14
	STO 3	33 03
	h RTN	35 22
	f LBL E	31 25 15
	STO 2	33 02
	h RTN	35 22
110	g LBLf a	32 25 11
	f CL REG	31 43
	h RTN	35 22

SP-D-002 - Slurry Charactersitics II

Input:

(1) solid concentration by weight (optional) - %
(2) slurry density (optional) - n (ratio)
(3) solid density - n (ratio)
(4) liquid density - n (ratio)
(5) suspension viscosity - centipoise
(6) solid specific heat - BTU/lb.°F
(7) liquid specific heat - BTU/lb.°F
(8) liquid thermal conductivity - BTU/hr.ft.°F
(9) solid thermal conductivity - BTU/hr.ft.°F

Computes:

(1) percent solids by volume - %
(2) solid volume fraction - n
(3) solid viscosity to suspension viscosity ratio - n
(4) slurry viscosity (Newtonian) - centipoise
(5) specific heat of slurry - BTU/lb.°F
(6) thermal conductivity of slurry - BTU/hr.ft.°F

STEP	INSTRUCTIONS	KEYS	OUTPUT UNITS
1.	load card SP-D-002, side 1 & 2		
2.	input solid density	C	n (ratio)
3.	input liquid density	D	n (ratio)
4.	input/compute solids concentration by weight	A	n%
5.	input/compute slurry density	B	n (ratio)
6.	input suspension viscosity	fa	centipoise
7.	input solid specific heat	fb	BTU/lb.°F
8.	input liquid specific heat	fc	BTU/lb.°F
9.	input liquid thermal conductivity	fd	BTU/hr.ft.°F
10.	input solid thermal conductivity	fe	BTU/hr.ft.°F
11.	Compute:	E	
	(1) percent solids by volume		n%
	(2) solid volume fraction		n
	(3) solid viscosity to suspension viscosity ratio		n
	(4) slurry viscosity (Newtonian)		centipoise
	(5) specific heat of slurry		BTU/lb.°F
	(6) thermal conductivity of slurry		BTU/hr.ft.°F
12.	New case desired: (clears and resets for a new case)	fa	

Example: SP-D-002

Given Conditions: Coal Concentrate Slurry

C_w: 48.28%

ρ_s: 1.4
ρ_l: 1.0
μ_o: 1.0 cp.
$C\rho_s$: .3 BTU/lb.°F
$C\rho_l$: 1.0 BTU/lb.°F
K_l: .349 (water @ 70°F) BTU/hr.ft.°F
K_s: .250 BTU/hr.ft.°F

Execution: Load card SP-D-002, side 1 & 2

KEYSTROKES	OUTPUT
1.4 "C"	1.400
1.0 "D"	1.000
48.28 "A"	48.280%
Press "B"	1.160 (computed slurry density)
1.0 "fa"	1.000 (liquid viscosity - cp.)
.3 "fb"	.300 (specific heat of solid)
1.0 "fc"	1.000 (specific heat of liquid)
.349 "fd"	.349 (thermal conductivity of liquid)
.250 "fe"	.250 (thermal conductivity of solid)

To Compute:

Press "E"	"40.004" (percent solids by volume)
	".4" (solids volume fraction)
	"5.699" (μ_m/μ_o ratio)
	"5.699" cp. (slurry viscosity - Newtonian suspension)
	".662" BTU/lb.°F (specific heat of slurry)
	".307" BTU/hr.ft.°F (thermal conductivity of slurry)

Do you wish to run a new case?
Press "fa"

STEP	KEY ENTRY	KEY CODE
001	f LBL A	31 25 11
	STO A	33 11
	h F? 3	35 71 03
	h RTN	35 22
	RCL B	34 12
	RCL 4	34 04
	-	51
	h ST I	35 33
	RCL 3	34 03
010	RCL 4	34 04
	-	51
	h RC I	35 34
	h x≷y	35 52
	÷	81
	RCL 3	34 03
	x	71
	RCL B	34 12
	÷	81
	1	01
020	0	00
	0	00
	x	71
	STO A	33 11
	h RTN	35 22
	f LBL B	31 25 12
	STO B	33 12
	h F? 3	35 71 03
	h RTN	35 22
	1	01
030	0	00
	0	00
	RCL A	34 11
	-	51
	RCL 4	34 04
	÷	81
	h ST I	35 33
	RCL A	34 11
	RCL 3	34 03
	÷	81
040	h RC I	35 34
	+	61
	1	01
	0	00
	0	00
	h x≷y	35 52
	÷	81
	STO B	33 12
	h RTN	35 22
	f LBL C	31 25 13
050	STO 3	33 03
	h RTN	35 22
	f LBL D	31 25 14
	STO 4	33 04
	h RTN	35 22
	f LBL E	31 25 15
	RCL A	34 11
	RCL B	34 12
	x	71
	RCL 3	34 03
060	÷	81
	STO C	33 13
	h PAUSE	35 72
	RCL A	34 11
	RCL B	34 12
	x	71
	1	01
	0	00
	0	00
	÷	81
070	RCL 3	34 03
	÷	81
	STO D	33 14
	h PAUSE	35 72
	1	01
	6	06
	.	83
	6	06
	x	71
	g e^x	32 52
080	.	83
	0	00
	0	00
	2	02
	7	07
	3	03
	x	71
	STO 9	33 09
	RCL D	34 14
	g x^2	32 54
090	1	01
	0	00
	.	83
	0	00
	5	05
	x	71
	STO +9	33 61 09
	RCL D	34 14
	2	02
	.	83
100	5	05
	x	71
	STO +9	33 61 09
	1	01
	STO +9	33 61 09
	RCL 9	34 09
	STO E	33 15
	0	00
	STO 9	33 09
	RCL E	34 15
110	h PAUSE	35 72
	RCL 5	34 05
	x	71

SP-D-002

STEP	KEY ENTRY	KEY CODE
	h PAUSE	35 72
	RCL 2	34 02
	RCL A	34 11
	x	71
	RCL A	34 11
	1	01
	0	00
120	0	00
	h x≷y	35 52
	-	51
	RCL 1	34 01
	x	71
	+	61
	1	01
	0	00
	0	00
	÷	81
130	h PAUSE	35 72
	RCL 7	34 07
	RCL 0	34 00
	h x≷y	35 52
	-	51
	RCL D	34 14
	x	71
	2	02
	x	71
	CHS	42
140	RCL 7	34 07
	+	61
	RCL 0	34 00
	2	02
	x	71
	+	61
	STO 6	33 06
	RCL 7	34 07
	RCL 0	34 00
	h x≷y	35 52
150	-	51
	RCL D	34 14
	x	71
	RCL 7	34 07
	+	61
	RCL 0	34 00
	2	02
	x	71
	+	61
	RCL 6	34 06
160	h x≷y	35 52
	÷	81
	RCL 0	34 00
	x	71
	h RTN	35 22
	g LBLf a	32 25 11
	STO 5	33 05
	h RTN	35 22
	g LBLf b	32 25 12
	STO 2	33 02
170	g LBLf c	32 25 13
	STO 1	33 01
	h RTN	35 22
	g LBLf d	32 25 14
	STO 0	33 00
	h RTN	35 22
	g LBLf e	32 25 15
	STO 7	33 07
	h RTN	35 22

SP-D-003 - Settling Velocity & Drag Coefficients Computation: Sphere

Input:

(1) solid density - lb./ft.^{3}
(2) liquid density - lb./ft.^{3}
(3) acceleration due to gravity - ft./sec.^{2}
(4) particle diameter - microns
(5) kinematic viscosity of fluid - ft.^{2}/sec.

Computes:

(1) type of flow, ie.
1 = Stokes law
2 = intermediate law
3 = Newton's law
(2) settling velocity - ft./sec.
(3) drag coefficient - n

STEP	INSTRUCTIONS	KEYS	OUTPUT UNITS
1.	load card SP-D-003, side 1 & 2		
2.	input solid density	A	lb./ft.^{3}
3.	input liquid density	A	lb./ft.^{3}
4.	input acceleration due to gravity	A	ft./sec.^{2}
5.	input particle diameter	A	microns
6.	input kinematic viscosity	A	ft.^{2}/sec.
7.	Compute:	B	
	(1) type of flow, ie.		n
	1 = Stokes law		
	2 = intermediate law		
	3 = Newton's law		
	(2) settling velocity		ft./sec.
	(3) drag coefficient		n
8.	Option to select a new particle diameter	fb	microns
9.	Option to clear for a new case	fa	
10.	Option to recall value of 1 centistoke in ft.^{2}/sec. units	fc	ft.^{2}/sec.

Example: SP-D-003

Given Conditions:

solid density: 87.22 lb./ft.^{3}
liquid density: 62.3 lb./ft.^{3}
acceleration due to gravity: 32.2 ft./sec.^{2}
particle diameter: 300 microns (sphere)
kinematic viscosity: 10 centistokes

Execution: Load card SP-D-003, side 1 & 2

Store parameters in this order:

KEYSTROKES	OUTPUT
87.22 "A"	87.22
62.30 "A"	62.30
32.2 "A"	32.20
300 "A"	300.00
"fc"	1.076×10^{-5} (ft.^{2}/sec.)
10 "x"	1.076×10^{-4} (ft.^{2}/sec.)
"A"	1.076×10^{-4} (ft.^{2}/sec.)
"DSP 5"	(will display 5 digit format)

To Compute:

Press "B"	"1.0" (Stokes law applies)
	".00644" (settling velocity)
	"407.262" (drag coefficient)

STEP	KEY ENTRY	KEY CODE
001	f LBL A	31 25 11
	STO (i)	33 24
	f GSB 1	31 22 01
	h RTN	35 22
	f LBL 1	31 25 01
	f ISZ	31 34
	R/S	84
	h ST I	35 33
	GTO 1	22 01
010	h RTN	35 22
	f LBL B	31 25 12
	f GSB E	31 22 15
	f GSB 2	31 22 02
	1	01
	.	83
	0	00
	g $x \leq y$	32 71
	GTO C	22 13
	GTO 3	22 03
020	h RTN	35 22
	f LBL E	31 25 15
	RCL 0	34 00
	RCL 1	34 01
	−	51
	RCL 3	34 03
	1	01
	0	00
	0	00
	0	00
030	÷	81
	2	02
	5	05
	.	83
	4	04
	÷	81
	1	01
	2	02
	÷	81
	g x^2	32 54
040	x	71
	RCL 2	34 02
	x	71
	1	01
	8	08
	÷	81
	RCL 1	34 01
	÷	81
	RCL 4	34 04
	÷	81
050	STO 5	33 05
	h RTN	35 22
	f LBL C	31 25 13
	RCL 0	34 00
	RCL 1	34 01
	−	51
	3	03

STEP	KEY ENTRY	KEY CODE
	.	83
	3	03
	3	03
060	x	71
	RCL 3	34 03
	1	01
	0	00
	0	00
	0	00
	÷	81
	2	02
	5	05
	.	83
070	4	04
	÷	81
	1	01
	2	02
	÷	81
	x	71
	RCL 1	34 01
	÷	81
	f $\sqrt{x}$	31 54
	STO 5	33 05
080	f GSB 2	31 22 02
	1	01
	0	00
	0	00
	0	00
	g $x \leq y$	32 71
	GTO 4	22 04
	GTO D	22 14
	h RTN	35 22
	f LBL D	31 25 14
090	RCL 0	34 00
	RCL 1	34 01
	−	51
	RCL 2	34 02
	x	71
	4	04
	x	71
	RCL 3	34 03
	1	01
	0	00
100	0	00
	0	00
	÷	81
	2	02
	5	05
	.	83
	4	04
	÷	81
	1	01
	2	02
110	÷	81
	3	03
	h y^x	35 63

STEP	KEY ENTRY	KEY CODE
	x	71
	RCL 1	34 01
	÷	81
	RCL 4	34 04
	g x^2	32 54
	÷	81
	3	03
120	÷	81
	STO 7	33 07
	.	83
	6	06
	9	09
	h y^x	35 63
	.	83
	1	01
	5	05
	x	71
130	STO 6	33 06
	RCL 4	34 04
	x	71
	RCL 3	34 03
	1	01
	0	00
	0	00
	0	00
	÷	81
	2	02
140	5	05
	.	83
	4	04
	÷	81
	1	01
	2	02
	÷	81
	÷	81
	STO 5	33 05
	GTO 5	22 05
150	h RTN	35 22
	f LBL 2	31 25 02
	RCL 3	34 03
	1	01
	0	00
	0	00
	0	00
	÷	81
	2	02
	5	05
160	.	83
	4	04
	÷	81
	1	01
	2	02
	÷	81
	RCL 5	34 05
	x	71
	RCL 4	34 04

STEP	KEY ENTRY	KEY CODE
	÷	81
170	STO 6	33 06
	h RTN	35 22
	f LBL 3	31 25 13
	2	02
	4	04
	RCL 6	34 06
	÷	81
	STO 8	33 08
	1	01
	f -x-	31 84
180	RCL 5	34 05
	f -x-	31 84
	RCL 8	34 08
	h RTN	35 22
	f LBL 4	31 25 04
	.	83
	4	04
	STO 8	33 08
	3	03
	f -x-	31 84
190	RCL 5	34 05
	f -x-	31 84
	RCL 8	34 08
	h RTN	35 22
	f LBL 5	31 25 05
	RCL 7	34 07
	RCL 6	34 06
	g x^2	32 54
	÷	81
	STO 8	33 08
200	2	02
	f -x-	31 84
	RCL 5	34 05
	f -x-	31 84
	RCL 8	34 08
	h RTN	35 22
	g LBLf a	32 25 11
	f CL REG	31 43
	h RTN	35 22
	g LBLf b	32 25 12
210	STO 3	33 03
	GTO B	22 12
	h RTN	35 22
	g LBLf c	32 25 13
	1	01
	.	83
	0	00
	7	07
	6	06
	EEX	43
220	CHS	42
	5	05
	ENTER ↑	41
	h RTN	35 22

SP-D-004 - Ratio of Volumetric concentration of Solids at 0.08D from top to pipe center. D = pipe diameter.

Input:

(1) velocity of slurry in the pipeline - ft./sec.
(2) viscosity of slurry - centipoise
(3) density of slurry - n (ratio)
(4) internal diameter of pipeline - inches
(5) length of pipeline - feet
(6) pressure drop per L - PSI/L
(7) settling velocity of particle - ft./sec.

Computes:

(1) Reynolds No. modified
(2) C/C_A ratio
(3) friction factor (optional)
(4) pressure drop per mile (optional)
(5) flow velocity (optional)

STEP	INSTRUCTIONS	KEYS	OUTPUT UNITS
1.	load card SP-D-004, side 1 & 2		
2.	Are the velocity and pressure drop per L given?		
	If YES - go to step 3		
	If NO - go to step 12 & 13		
3.	input flow velocity	A	ft./sec.
4.	input slurry viscosity	A	centipoise
5.	input slurry density	A	n (ratio)
6.	input internal diameter of pipe	A	inches
7.	input length of pipeline	A	feet
8.	input pressure drop per L in step 7	A	PSI/L
9.	input settling velocity of particle	A	feet/sec.
10.	Compute:	B	
	(1) Reynolds no. modified		n
	(2) C/C_A ratio		n
11.	Optional:		
	(1) input new settling velocity	fc	ft./sec.
	(2) Compute:	B	
	(a) Reynolds no.		n
	(b) C/C_A ratio		n
12.	Compute flow velocity: (OPTIONAL)		
	(1) enter slurry viscosity	STO 1	cp
	(2) enter slurry density	STO 2	n (ratio)
	(3) enter I.D.	STO 3	inches
	(4) enter pipeline length	STO 4	feet
	(5) enter settling velocity of particle	STO 6	ft./sec.

STEP	INSTRUCTIONS	KEYS	OUTPUT UNITS
	(6) enter slurry flowrate in gpm	ENTER↑	gpm
	(7) compute flow velocity	C	ft./sec.
13.	Compute pressure drop per L ft.:	D	
	(1) Reynolds no. modified		n
	(2) friction factor		n
	(3) pressure drop per L		PSI/L

Note: After velocity and pressure drop have been computed, go to step 3 and continue sequential execution.

STEP	INSTRUCTIONS	KEYS	OUTPUT UNITS
14.	Optional Information:		
	(1) C/C_A ratio	RCL A	n
	(2) pressure drop per L	RCL B	PSI/L
	(3) flow velocity	RCL C	ft./sec.
	(4) Reynolds no. modified	RCL D	n
	(5) friction factor	RCL E	n
15.	To clear for new case:	fa	

Example: SP-D-004, C/C_A Ratio Computation

Given Conditions:

slurry flowrate: 1,589 gpm
slurry viscosity: 7.9 centipoise
slurry density: 1.92
internal diameter: 12 inches
length of pipeline: 5,280 feet
settling velocity of particles: .00858 ft./sec.

Execution: Load card SP-D-004, side 1 & 2

Since flow velocity and pressure drop are not given, execute in the following sequence:

KEYSTROKES	OUTPUT
7.9 "STO 1"	7.90000
1.92 "STO 2"	1.92000
12 "STO 3"	12.00000
5,280 "STO 4"	5,280.00000
.00858 "STO 6"	.00858

To Compute:

KEYSTROKES	OUTPUT
1589 "C"	4.5 ft./sec. (flow velocity)
Press "D"	"13.13" (Reynolds no.)
	".00445" (friction factor)
	"24.63" PSI/mile

Go to step 3 of instructions.

Store parameters in the following order:

KEYSTROKES	OUTPUT
4.5 "A"	4.50000
7.9 "A"	7.90000
1.92 "A"	1.92000
12 "A"	12.00000
5,280 "A"	5,280.00000
24.63 "A"	24.63000
.00858 "A"	.00858

To Compute:

Press "B"	"13.12" (Reynolds no.)
	".64373" (C/C_A)
RCL A	.64373 (C/C_A ratio)
RCL B	24.633 PSI/mi.
RCL C	4.5 ft./sec.
RCL D	13.13
RCL E	.00445

Option to input a new settling velocity:

.00242 "fc"	.00242
Press "B"	"13.12" (Reynolds No. modified)
	".88316" (C/C_A ratio)
.00031 "fc"	.00031
Press "B"	"13.12" (Reynolds no. modified)
	".98421" (C/C_A ratio)

To clear for a new case:
"fa"

STEP	KEY ENTRY	KEY CODE
001	f LBL A	31 25 11
	STO (i)	33 24
	f GSB 1	31 22 01
	h RTN	35 22
	f LBL 1	31 25 01
	f ISZ	31 34
	R/S	84
	h ST I	35 33
	GTO 1	22 01
010	h RTN	35 22
	f LBL B	31 25 12
	RCL 6	34 06
	1	01
	.	83
	8	08
	x	71
	.	83
	3	03
	8	08
020	÷	81
	STO A	33 11
	f GSB 2	31 22 02
	RCL A	34 11
	h x≷y	35 52
	÷	81
	CHS	42
	g 10^x	32 53
	STO A	33 11
	h RTN	35 22
030	f LBL 2	31 25 02
	RCL 3	34 03
	RCL 2	34 02
	RCL 0	34 00
	x	71
	x	71
	RCL 1	34 01
	÷	81
	STO 7	33 07
	h PAUSE	35 72
040	RCL 5	34 05
	RCL 3	34 03
	x	71
	.	83
	3	03
	2	02
	3	03
	÷	81
	RCL 4	34 04
	÷	81
050	RCL 2	34 02
	÷	81
	RCL 0	34 00
	g x^2	32 54
	÷	81
	STO 8	33 08
	2	02
	÷	81
	f $\sqrt{x}$	31 54
	RCL 0	34 00
060	x	71
	STO 9	33 09
	h RTN	35 22
	g LBLf a	32 25 11
	f CL REG	31 43
	h RTN	35 22
	f LBL C	31 25 13
	STO C	33 13
	4	04
	2	02
070	÷	81
	6	06
	0	00
	x	71
	2	02
	4	04
	x	71
	.	83
	0	00
	1	01
080	1	01
	9	09
	x	71
	RCL 3	34 03
	g x^2	32 54
	÷	81
	STO C	33 13
	h RTN	35 22
	f LBL D	31 25 14
	RCL 3	34 03
090	RCL C	34 13
	RCL 2	34 02
	x	71
	x	71
	RCL 1	34 01
	÷	81
	STO D	33 14
	h PAUSE	35 72
	.	83
	4	04
100	g x≤y	32 71
	GTO 7	22 07
	GTO 8	22 08
	h RTN	35 22
	f LBL 7	31 25 07
	RCL D	34 14
	h 1/x	35 62
	.	83
	3	03
	5	05
110	5	05
	h y^x	35 63
	.	83

STEP	KEY ENTRY	KEY CODE
	0	00
	0	00
	6	06
	6	06
	2	02
	x	71
	.	83
120	0	00
	0	00
	1	01
	8	08
	+	61
	STO E	33 15
	h PAUSE	35 72
	f GSB 5	31 22 05
	RCL E	34 15
	x	71
130	STO B	33 12
	h RTN	35 22
	f LBL 8	31 25 08
	RCL D	34 14
	h 1/x	35 62
	.	83
	0	00
	0	00
	2	02
	0	00
140	7	07
	x	71
	STO E	33 15
	h PAUSE	35 72
	f GSB 5	31 22 05
	RCL E	34 15
	x	71
	STO B	33 12
	h RTN	35 22
	f LBL 5	31 25 05
150	RCL C	34 13
	g x^2	32 54
	RCL 2	34 02
	RCL 4	34 04
	x	71
	x	71
	.	83
	3	03
	2	02
	3	03
160	x	71
	RCL 3	34 03
	÷	81
	h RTN	35 22
	g LBLf c	32 25 13
	STO 6	33 06
	h RTN	35 22

SP-D-005 - Homogeneous Flow, $C/C_A \geq .8$, Computation

Input:

(1) solids percent by weight - n%
(2) internal diameter of pipeline - inches
(3) plastic viscosity - centipoise
(4) yield stress - $dynes/cm.^2$
(5) flowrate velocity - ft./sec.
(6) density of solids - n (ratio)
(7) density of liquid - n (ratio)

Computes:

(1) transition velocity - ft./sec.
(2) pressure drop per mile - PSI/mi.

STEP	INSTRUCTIONS	KEYS	OUTPUT UNITS
1.	load card SP-D-005, side 1 & 2		
2.	input solids percent by weight	A	n%
3.	input internal diameter	A	inches
4.	input plastic viscosity	A	centipoise
5.	input yield stress	A	$dynes/cm.^2$
6.	input flow velocity	A	ft./sec.
7.	input density of solids	A	n (ratio)
8.	input density of fluids	A	n (ratio)
9.	Compute:	B	
	(1) transition velocity		ft./sec.
	(2) pressure drop per mile		PSI/mi.

Example #1: SP-D-005

Given Conditions:

C_w: 65%
I.D.: 6 inches
μ: 22 cp.
τ_o: 53 $dynes/cm.^2$
v: 4 ft./sec.
ρ_s: 2.7
ρ_l: 1.0

Execution: Load card SP-D-005, side 1 & 2

Store parameters in this order:

KEYSTROKES	OUTPUT
65 "A"	65.00
6 "A"	6.00
22 "A"	22.00

KEYSTROKES	OUTPUT
53 "A"	53.00
4 "A"	4.00
2.7 "A"	2.70
1.0 "A"	1.00

To Compute:

Press "B"	"3.88" ft./sec. (velocity of transition)
	"54.90" PSI/mi. (pressure drop per mile)

Optional:

RCL 7	14,246.11 (Reynolds No.)
RCL 8	13,831.01 (critical Reynolds No.)
f P≷S	
RCL 3	427,633.74 (Hedstrom No.)
RCL 5	.00713 (friction factor) - DSP 5
f P≷S	

If new case is desired:
"fa"

Example #2: SP-D-005

Given Conditions:

C_w: 50%
I.D.: 8 inches
viscosity: 22 cp.
yield stress: 200 dynes/cm.^{2}
flow velocity: 4.0 ft./sec.
solid density: 2.7
liquid density: 1.0

Store parameters in this order:

KEYSTROKES	OUTPUT
50 "A"	50.00
8 "A"	8.00
22 "A"	22.00
200 "A"	200.00
4 "A"	4.00
2.7 "A"	2.70
1.0 "A"	1.00

To Compute:

Press "B" ____________ "5.82" ft./sec. (velocity of transition)

Flashing denotes flow velocity is critical.
Recall original flow velocity from "Given Conditions" above = 4.00 ft./sec.

Since the flow velocity is under the transition velocity, particles will settle to the bottom of the pipe.

STEP	KEY ENTRY	KEY CODE
001	f LBL A	31 25 11
	STO (i)	33 24
	f GSB 1	31 22 01
	h RTN	35 22
	f LBL 1	31 25 01
	f ISZ	31 34
	R/S	84
	h ST I	35 33
	GTO 1	22 01
010	h RTN	35 22
	f LBL 2	31 25 02
	RCL 0	34 00
	RCL 5	34 05
	÷	81
	1	01
	0	00
	0	00
	RCL 0	34 00
	CHS	42
020	+	61
	RCL 6	34 06
	÷	81
	+	61
	1	01
	0	00
	0	00
	h x≷y	35 52
	÷	81
	3	03
030	2	02
	.	83
	2	02
	÷	81
	RCL 6	34 06
	6	06
	2	02
	.	83
	3	03
	x	71
040	x	71
	f P≷S	31 42
	STO 2	33 02
	f P≷S	31 42
	h RTN	35 22
	f LBL 3	31 25 03
	RCL 1	34 01
	1	01
	2	02
	÷	81
050	RCL 4	34 04
	f P≷S	31 42
	RCL 2	34 02
	f P≷S	31 42
	x	71
	x	71
	RCL 2	34 02
	÷	81
	2	02
	.	83
060	0	00
	9	09
	EEX	43
	CHS	42
	5	05
	ENTER ↑	41
	h R↓	35 53
	÷	81
	STO 7	33 07
	h RTN	35 22
070	f LBL 4	31 25 04
	RCL 3	34 03
	4	04
	8	08
	0	00
	÷	81
	f P≷S	31 42
	RCL 2	34 02
	f P≷S	31 42
	x	71
080	RCL 1	34 01
	1	01
	2	02
	÷	81
	g x^2	32 54
	x	71
	RCL 2	34 02
	g x^2	32 54
	÷	81
	2	02
090	.	83
	0	00
	9	09
	EEX	43
	CHS	42
	5	05
	ENTER ↑	41
	h R↓	35 53
	g x^2	32 54
	÷	81
100	f P≷S	31 42
	STO 3	33 03
	f P≷S	31 42
	h RTN	35 22
	f LBL 5	31 25 05
	f P≷S	31 42
	RCL 3	34 03
	f P≷S	31 42
	.	83
	3	03
110	1	01
	h y^x	35 63
	2	02

STEP	KEY ENTRY	KEY CODE
	4	04
	8	08
	.	83
	4	04
	4	04
	x	71
	STO 8	33 08
120	h RTN	35 22
	f LBL 6	31 25 06
	RCL 8	34 08
	RCL 2	34 02
	x	71
	2	02
	.	83
	0	00
	9	09
	EEX	43
130	CHS	42
	5	05
	ENTER ↑	41
	h R↓	35 53
	x	71
	RCL 1	34 01
	1	01
	2	02
	÷	81
	÷	81
140	f P≷S	31 42
	RCL 2	34 02
	÷	81
	STO 4	33 04
	f P≷S	31 42
	h RTN	35 22
	f LBL 7	31 25 07
	RCL 7	34 07
	RCL 8	34 08
	g x≤y	32 71
150	GTO 8	22 08
	GTO 9	22 09
	h RTN	35 22
	f LBL 8	31 25 08
	RCL 7	34 07
	7	07
	7	07
	4	04
	2	02
	÷	81
160	h 1/x	35 62
	.	83
	3	03
	5	05
	5	05
	h y^x	35 63
	.	83
	0	00
	0	00
	6	06
170	6	06
	2	02
	x	71
	.	83
	0	00
	0	00
	1	01
	8	08
	+	61
	f P≷S	31 42
180	STO 5	33 05
	RCL 2	34 02
	8	08
	8	08
	1	01
	.	83
	4	04
	6	06
	x	71
	RCL 5	34 05
190	x	71
	f P≷S	31 42
	RCL 4	34 04
	f P≷S	31 42
	g x^2	32 54
	x	71
	f P≷S	31 42
	RCL 1	34 01
	÷	81
	f P≷S	31 42
200	STO 1	33 01
	f P≷S	31 42
	h RTN	35 22
	f LBL 9	31 25 09
	f P≷S	31 42
	RCL 4	34 04
	f -x-	31 84
	f P≷S	31 42
	RCL 4	34 04
	f -x-	31 84
210	h RTN	35 22
	g LBLf a	32 25 11
	f CL REG	31 43
	f P≷S	31 42
	f CL REG	31 43
	h RTN	35 22
	f LBL B	31 25 12
	f GSB 2	31 22 02
	f GSB 3	31 22 03
	f GSB 4	31 22 04
220	f GSB 5	31 22 05
	f GSB 6	31 22 06
	h PAUSE	35 72
	f GSB 7	31 22 07
	h RTN	35 22

SP-D-006A - Vehicle Friction Loss Computation

Input:

(1) solid density - n (ratio)
(2) liquid density - n (ratio)
(3) internal diameter - inches
(4) flowrate velocity - ft./sec.
(5) C_v (volume percent) - n%
(6) viscosity - centipoise

Computes:

(1) Reynolds No.
(2) friction factor
(3) friction velocity
(4) pressure drop per mile for slurry
(5) pressure drop - ft. of slurry/ft. - slurry
(6) pressure drop - ft. of water/ft. - slurry

STEP	INSTRUCTIONS	KEYS	OUTPUT UNITS
1.	load card SP-D-006A, side 1 & 2		
2.	input solid density	A	n (ratio)
3.	input liquid density	A	n (ratio)
4.	input internal diameter	A	inches
5.	input flow velocity	A	ft./sec.
6.	input solid volume percent	A	n
7.	input viscosity of slurry @ volume percent	A	centipoise
8.	Compute: (slurry)	B	
	(1) Reynolds No.		n
	(2) friction factor		n
	(3) friction velocity		ft./sec.
	(4) pressure drop per mile for vehicle		PSI/mi.
9.	Compute: (slurry)	D	
	(1) slurry pressure drop		ft. slurry/ft.
	(2) slurry pressure drop		ft. water/ft.
10.	Optional: (Water pressure drop computation)		
	(1) Compute: (water)	C	
	(1) Reynolds No.		n
	(2) friction factor		n
	(3) friction velocity		ft./sec.
	(4) pressure drop per mile		PSI/mi.
	(2) Compute: (water) Water pressure drop	E	ft. water/ft.
11.	Optional: New case desired? Clear all primary registers for new input	fa	

Example: SP-D-006A - Vehicle Pressure Drop Computation

Given Conditions:

solid density: 5.0 (iron ore)
liquid density: 1.0
internal diameter: 12 inches
flow velocity: 4.5 ft./sec.
solids percent by volume: 23%
slurry viscosity @ 23% volume concentration: 7.9 cp.

Execution: Load card SP-D-006A, side 1 & 2

Store parameters in this order:

KEYSTROKES	OUTPUT
5.0 "A"	5.00
1.0 "A"	1.00
12 "A"	12.00
4.5 "A"	4.50
23 "A"	23.00
7.9 "A"	7.90

To Compute:

Slurry - Press "B"	"101,392.41" (Reynolds No.)
	".0045" (friction factor)
	".2124" (friction velocity)
	"24.6236" (pressure drop per mile)
Slurry - Press "D"	".0056" ft. slurry/ft. (slurry pressure drop)
	".0108" ft. water/ft. (slurry pressure drop)
Water - Press "C"	"417,187.5" (Reynolds No.)
	".0034" (friction factor)
	".1857" (friction velocity)
	"9.8069" (pressure drop per mile)
Water - Press "E"	.0043 ft. water/ft. (pressure drop)

New case desired?
To Clear Registers:
"fa"

STEP	KEY ENTRY	KEY CODE
001	f LBL A	31 25 11
	STO (i)	33 24
	f GSB 1	31 22 01
	h RTN	35 22
	f LBL 1	31 25 01
	f ISZ	31 34
	R/S	84
	h ST I	35 33
	GTO 1	22 01
010	h RTN	35 22
	f LBL B	31 25 12
	RCL 2	34 02
	1	01
	2	02
	÷	81
	RCL 3	34 03
	x	71
	RCL 4	34 04
	1	01
020	0	00
	0	00
	÷	81
	RCL 0	34 00
	1	01
	-	51
	x	71
	1	01
	+	61
	STO E	33 15
030	6	06
	2	02
	.	83
	3	03
	x	71
	x	71
	RCL 5	34 05
	÷	81
	6	06
	.	83
040	7	07
	2	02
	EEX	43
	CHS	42
	4	04
	ENTER ↑	41
	h R↓	35 53
	÷	81
	STO 6	33 06
	h PAUSE	35 72
050	7	07
	7	07
	4	04
	2	02
	÷	81
	h 1/x	35 62
	.	83

STEP	KEY ENTRY	KEY CODE
	3	03
	5	05
	5	05
060	h y^x	35 63
	.	83
	0	00
	0	00
	6	06
	6	06
	2	02
	x	71
	.	83
	0	00
070	0	00
	1	01
	8	08
	+	61
	h PAUSE	35 72
	STO 7	33 07
	2	02
	÷	81
	f $\sqrt{x}$	31 54
	RCL 3	34 03
080	x	71
	h PAUSE	35 72
	STO 8	33 08
	RCL E	34 15
	RCL 7	34 07
	RCL 3	34 03
	g x^2	32 54
	x	71
	x	71
	1	01
090	7	07
	0	00
	5	05
	.	83
	4	04
	4	04
	x	71
	RCL 2	34 02
	÷	81
	STO 9	33 09
100	h RTN	35 22
	g LBLf a	32 25 11
	f CL REG	31 43
	h RTN	35 22
	f LBL C	31 25 13
	f P≷S	31 42
	1	01
	STO 0	33 00
	1	01
	STO 2	33 02
110	f P≷S	31 42
	RCL 2	34 02
	f P≷S	31 42

STEP	KEY ENTRY	KEY CODE
	STO 2	33 02
	f P≷S	31 42
	RCL 3	34 03
	f P≷S	31 42
	STO 3	33 03
	1	01
	0	00
120	0	00
	STO 4	33 04
	1	01
	STO 5	33 05
	f GSB B	31 22 12
	f P≷S	31 42
	h RTN	35 22
	f LBL D	31 25 14
	RCL 9	34 09
	5	05
130	2	02
	8	08
	0	00
	÷	81
	.	83
	4	04
	3	03
	3	03
	÷	81
	RCL 0	34 00
140	RCL 1	34 01
	CHS	42
	+	61
	RCL 4	34 04
	1	01
	0	00
	0	00
	÷	81
	x	71
	1	01
150	+	61
	STO A	33 11
	÷	81
	h PAUSE	35 72
	RCL A	34 11
	x	71
	h RTN	35 22
	f LBL E	31 25 15
	f P≷S	31 42
	RCL 9	34 09
160	.	83
	4	04
	3	03
	3	03
	÷	81
	5	05
	2	02
	8	08
	0	00
	÷	81
170	f P≷S	31 42
	h RTN	35 22

SP-D-006B - Vehicle & Bed Volume Fraction Computation

Input:

(1) % weight for diameter 1 - n
(2) % weight for diameter 2 - n
(3) % weight for diameter 3 - n
(4) % weight for diameter 4 - n
(5) % weight for diameter 5 - n
(6) C/C_A ratio for diameter 1 - n
(7) C/C_A ratio for diameter 2 - n
(8) C/C_A ratio for diameter 3 - n
(9) C/C_A ratio for diameter 4 - n
(10) C/C_A ratio for diameter 5 - n
(11) total volume percent of solids - n

Computes:

(1) vehicle volume fraction, d_1 - n
bed volume fraction, d_1 - n
(2) vehicle volume fraction, d_2 - n
bed volume fraction, d_2 - n
(3) vehicle volume fraction, d_3 - n
bed volume fraction, d_3 - n
(4) vehicle volume fraction, d_4 - n
bed volume fraction, d_4 - n
(5) vehicle volume fraction, d_5 - n
bed volume fraction, d_5 - n
(6) summation of vehicle volume fractions - n

STEP	INSTRUCTIONS	KEYS	OUTPUT UNITS
1.	load card SP-D-006B, side 1 & 2		
2.	input total solid volume percent	fb	n
3.	input % weight for d_1	A	n
	$-d_2$	A	n
	$-d_3$	A	n
	$-d_4$	A	n
	$-d_5$	A	n
4.	input C/C_A ratio for d_1	A	n
	$-d_2$	A	n
	$-d_3$	A	n
	$-d_4$	A	n
	$-d_5$	A	n
5.	Compute:	B	
	(1) vehicle solid volume fraction-d_1		n
	(2) bed solid volume fraction-d_1		n
	(3) vehicle solid volume fraction-d_2		n
	(4) bed solid volume fraction-d_2		n
	(5) vehicle solid volume fraction-d_3		n

STEP	INSTRUCTIONS	KEYS	OUTPUT UNITS
	(6) bed solid volume fraction-d_3		n
	(7) vehicle solid volume fraction-d_4		n
	(8) bed solid volume fraction-d_4		n
	(9) vehicle solid volume fraction-d_5		n
	(10) bed solid volume fraction-d_5		n
6.	Compute summation of vehicle solid volume fraction	C	n
7.	New case	fa	resets program

Example: SP-D-006B

Given Conditions:

total solid volume percent: 23%

	Tyler mesh	% weight	C/C_A Ratio
d_1 -	100 x 200	5.0	.64312
d_2 -	200 x 325	10.0	.8827
d_3 -	-325	85.0	.9842
d_4 -	0	0	0
d_5 -	0	0	0

Execution: Load card SP-D-006B, side 1 & 2

Store parameters in this order:

KEYSTROKES	OUTPUT
23 "fb"	23.00000
5 "A"	5.00000
10 "A"	10.00000
85 "A"	85.00000
0 "A"	0.00000 (necessary input)
0 "A"	0.00000 (necessary input)
.64312 "A"	.64312
.8827 "A"	.88270
.9842 "A"	.98420

To Compute vehicle & bed solid volume fractions:

Press "B"	".73959" (vehicle) d_1
	".41041" (bed) d_1
	"2.03021" (vehicle) d_2
	".26979" (bed) d_2
	"19.24111 (vehicle) d_3
	".30890" (bed) d_3

KEYSTROKES	OUTPUT
___________	"0.00" (vehicle) d_4
___________	"0.00" (bed) d_4
___________	"0.00" (vehicle) d_5
___________	"0.00" (bed) d_5
Press "C" ___________	22.01 (summation of all vehicle volume fractions)

Optional:

To clear for new case:
"fa"_______________resets program

STEP	KEY ENTRY	KEY CODE
001	f LBL A	31 25 11
	STO (i)	33 24
	f GSB 1	31 22 01
	h RTN	35 22
	f LBL 1	31 25 01
	f ISZ	31 34
	R/S	84
	h ST I	35 33
	GTO 1	22 01
010	h RTN	35 22
	f LBL B	31 25 12
	RCL B	34 12
	RCL 0	34 00
	f %	31 82
	STO 0	33 00
	RCL 5	34 05
	x	71
	f P≷S	31 42
	STO 0	33 00
020	f P≷S	31 42
	h PAUSE	35 72
	RCL 0	34 00
	f P≷S	31 42
	RCL 0	34 00
	CHS	42
	+	61
	STO 5	33 05
	f P≷S	31 42
	h PAUSE	35 72
030	RCL B	34 12
	RCL 1	34 01
	f %	31 82
	STO 1	33 01
	RCL 6	34 06
	x	71
	f P≷S	31 42
	STO 1	33 01
	f P≷S	31 42
	h PAUSE	35 72
040	CHS	42
	RCL 1	34 01
	+	61
	f P≷S	31 42
	STO 6	33 06
	f P≷S	31 42
	h PAUSE	35 72
	RCL B	34 12
	RCL 2	34 02
	f %	31 82
050	STO 2	33 02
	RCL 7	34 07
	x	71
	f P≷S	31 42
	STO 2	33 02
	f P≷S	31 42
	h PAUSE	35 72
	CHS	42
	RCL 2	34 02
	+	61
060	f P≷S	31 42
	STO 7	33 07
	f P≷S	31 42
	h PAUSE	35 72
	RCL B	34 12
	RCL 3	34 03
	f %	31 82
	STO 3	33 03
	RCL 8	34 08
	x	71
070	f P≷S	31 42
	STO 3	33 03
	f P≷S	31 42
	h PAUSE	35 72
	CHS	42
	RCL 3	34 03
	+	61
	f P≷S	31 42
	STO 8	33 08
	f P≷S	31 42
080	h PAUSE	35 72
	RCL B	34 12
	RCL 4	34 04
	f %	31 82
	STO 4	33 04
	RCL 9	34 09
	x	71
	f P≷S	31 42
	STO 4	33 04
	f P≷S	31 42
090	h PAUSE	35 72
	CHS	42
	RCL 4	34 04
	+	61
	f P≷S	31 42
	STO 9	33 09
	f P≷S	31 42
	h RTN	35 22
	g LBLf a	32 25 11
	f CL REG	31 43
100	f P≷S	31 42
	f CL REG	31 43
	h RTN	35 22
	g LBLf b	32 25 12
	STO B	33 12
	h RTN	35 22
	f LBL C	31 25 13
	f P≷S	31 42
	RCL 0	34 00
	RCL 1	34 01
110	+	61
	RCL 2	34 02
	+	61

STEP	KEY ENTRY	KEY CODE
	RCL 3	34 03
	+	61
	RCL 4	34 04
	+	61
	f P$\gtrless$S	31 42
	h RTN	35 22

SP-D-007 - Slurry Pipeline Vehicle & Bed Pressure Drop Computation

Input:

(1) internal diameter of pipeline - inches
(2) acceleration due to gravity - ft./sec.^{2}
(3) flow velocity - ft./sec.
(4) drag coefficient - n
(5) volume fraction of solids (bed) - n
(6) pressure drop per mile (water) - PSI/mi.
(7) solid density - n (ratio)
(8) volume percent solids (vehicle) - n%
(9) viscosity of slurry @ C_v - centipoise
(10) liquid density - n (ratio)
(11) flowrate - gpm
(12) total pressure drop, vehicle & bed - PSI
(13) pump efficiency - n

Computes:

(1) bed pressure drop per mile - PSI/mi.
(2) Reynolds No. - vehicle - n
(3) friction factor - vehicle - n
(4) pressure drop per mile - vehicle - PSI/mi.

STEP	INSTRUCTIONS	KEYS	OUTPUT UNITS
1.	load card SP-D-007, side 1 & 2		
2.	input internal diameter	A	inches
3.	input acceleration due to gravity	A	ft./sec.^{2}
4.	input flow velocity	A	ft./sec.
5.	input drag coefficient using water settling velocity (d_n)	A	n
6.	input volume fraction for bed	A	n
7.	input pressure drop/mi. for water	A	PSI/mi.
8.	input solids density	A	n
9.	input volume percent solids	A	n%
10.	input slurry viscosity @ volume percent solids	A	centipoise
11.	input liquid density	A	n
12.	input slurry flowrate	A	gpm
13.	input total pressure drop (vehicle & bed)	A	PSI
14.	input pump efficiency	A	n
	Second Iteration		
15.	Compute: (bed) Bed pressure drop as a function of bed volume fraction	B	PSI/mi.

STEP	INSTRUCTIONS	KEYS	OUTPUT UNITS
16.	Compute: (vehicle)	C	
	(1) Reynolds No.		n
	(2) friction factor		n
	(3) pressure drop/mi.		PSI/mi.
17.	(1) input drag coefficient	D	n
	(2) input volume fraction for bed	E	n
	(3) compute vehicle pressure drop per mile	B	PSI/mi.
18.	(1) input drag coefficient	D	n
	(2) input volume fraction for bed	E	n
	(3) compute vehicle pressure drop per mile	B	PSI/mi.
19.	Compute summation of bed pressure drop and vehicle pressure drop	fc	PSI/mi.
20.	Optional:		
	(1) compute approximate flowrate of slurry	fd	gpm
	(2) Compute BHP:		
	(a) input flowrate	A	gpm
	(b) input total pressure required	A	PSI
	(c) input pump efficiency	A	n
	(d) compute BHP	fe	BHP
21.	To clear and reset for new case:	fa	

Example: SP-D-007 TWO PHASE VEHICLE COMPUTATION

Determine the friction loss for an iron concentrate slurry flowing in a 12 inch inside diameter horizontal pipe.

Given Conditions:

solid specific gravity: 5.0
solid shape factor: 1.0
solid size:

	Microns	Tyler Mesh	% Weight
d_1 -	(149 + 74).5 = 111.5	100 x 200	5.0
d_2 -	(74 + 44).5 = 59	200 x 325	10.0
d_3 -	(44 + 0).5 = 22	-325	85.0

liquid density: 1.0
solids concentration by volume: 23%
slurry temperature: 68°F

Slurry Viscosity Volume % Solids	Viscosity Centipoise
23	7.9
20	5.6
18	4.6
16	3.7
14	3.1

flow velocity: 4.5 ft./sec.; line length: 30 miles

Execution: Load card SP-D-006A, side 1 & 2: CLEAR ALL REGISTERS

Store parameters in this order:

KEYSTROKES	OUTPUT
5.0 "A"	5.00
1.0 "A"	1.00
12 "A"	12.00
4.5 "A"	4.50
23 "A"	23.00
7.9 "A"	7.90

To Compute:

Press "B" ________ "101,392.41" (Reynolds No.)
________ ".0045" (friction factor)
________ ".2124" (friction velocity)
________ "24.62" (pressure drop per mile)

Slurry - Press "D" ________ ".0056" ft. slurry/ft. (slurry pressure drop)
________ ".0108" ft. water/ft. (slurry pressure drop)

Water - Press "C" ________ "417,187.5" (Reynolds No.)
________ ".0034" (friction factor)
________ ".1857" (friction velocity)
________ "9.81" (pressure drop per mile)

Press "E" ________ .0043 ft. water/ft.

RCL A ________ 1.92 slurry density (ratio)

Load card SP-D-003, side 1 & 2: CLEAR ALL REGISTERS

Store parameters in this order:

62.3 x 5.0 "A" ________ 311.50
62.3 "A" ________ 62.30
32.2 "A" ________ 32.20
111 "A" ________ 111.00 (d_1 particle diameter)
"fc" "A" ________ 1.076×10^{-5}

To Compute parameters in water:

	KEYSTROKES	OUTPUT
	Press "B"	"2.0" (Intermediate law)
		".08446" ft./sec. (settling velocity)
		"8.77" (drag coefficient)
(d_2, particle diameter)	59 "fb"	"1.0" (Stokes law)
		".02442" ft./sec. (settling velocity)
		"53.3" (drag coefficient)
(d_3, particle diameter)	21 "fb"	"1.0" (Stokes law)
		".00316" ft./sec. (settling velocity)
		"1187.4" (drag coefficient)

Compute settling velocity of particles and drag coefficients:

"fa" ——— clear and reset program

Store parameters in this order: "CLEAR ALL REGISTERS"

	KEYSTROKES	OUTPUT
	5.0 x 62.3 "A"	311.500
	1.92 x 62.3 "A"	119.616
	32.2 "A"	32.200
	111 "A"	111.000 (d_1, particle diameter)
	$\frac{(\text{"fc"} \times 7.9)}{1.92}$ "A"	.000044273

To Compute parameters in slurry:

	KEYSTROKES	OUTPUT
	Press "B"	"1.0" (Stokes law)
		".008596" ft./sec. (settling velocity)
		"339.4" (drag coefficient)
(d_2, particle diameter)	59 "fb"	"1.0" (Stokes law)
		".00243" ft./sec. (settling velocity)
		"2,260" (drag coefficient)
(d_3, particle diameter)	21 "fb"	"1.0" (Stokes law)
		".00031" ft./sec. (settling velocity)
		"50,123" (drag coefficient)

Load card SP-D-004, side 1 & 2: "CLEAR ALL REGISTERS"

Store parameters in this order:

	KEYSTROKES	OUTPUT
	4.5 "A"	4.500000
	7.9 "A"	7.900000
	1.92 "A"	1.920000
	12 "A"	12.000000
	5,280 "A"	5,280.000000
	24.62 "A"	24.620000
(ω_1)	.008596 "A"	.008596

To Compute:

KEYSTROKES	OUTPUT
Press "B"	"13.12" (Reynolds No. modified)-Pipeline
	".64312" (C/C_A ratio)
(ω_2) .00243 "fc"	.00243
Press "B"	"13.12" (Reynolds No. modified)-Pipeline
	".8827" (C/C_A ratio)
(ω_3) .00031 "fc"	.00031
Press "B"	"13.12" (Reynolds No. modified)-Pipeline
	".9842" (C/C_A ratio)

Load card SP-D-006B, side 1 & 2: "CLEAR ALL REGISTERS"

Store parameters in this order:

23.0 "fb"	23.00000
5.0 "A"	5.00000
10.0 "A"	10.00000
85.0 "A"	85.00000
0.00 "A"	0.00000
0.00 "A"	0.00000
.64312 "A"	.64312
.8827 "A"	.88270
.9842 "A"	.98420
0.00 "A"	0.00000
0.00 "A"	0.00000

To Compute:

Press "B"	".73959" (vehicle fraction, d_1)
	".41041" (bed fraction, d_1)
	"2.03021" (vehicle fraction, d_2)
	".26979" (bed fraction, d_2)
	"19.24111" (vehicle fraction, d_3)
	".30890" (bed fraction, d_3)
	"0.00" (vehicle fraction, d_4)
	"0.00" (bed fraction, d_4)
	"0.00" (vehicle fraction, d_5)
	"0.00" (bed fraction, d_5)
Press "C"	22.01 (vehicle fraction summation)

Load card SP-D-007, side 1 & 2: "CLEAR ALL REGISTERS"

"f CL REG" ________ to reset program

Store parameters in this order:

12 "A" ________ 12.00

KEYSTROKES	OUTPUT
32.2 "A"	32.20
4.5 "A"	4.50
8.77 "A"	8.77
.41 "A"	.41 (bed fraction)
9.81 "A"	9.81
5 "A"	5.00
22.01 "A"	22.01 (vehicle fraction summation)
7 "A"	7.00 (viscosity @ 22.01% concentration)
1.0 "A"	1.00 (P_L)

To Compute:

Press "B"	10.38 PSI/mi. (bed pressure drop for d_1 particles)
Press "C"	"112,068.48" (Reynolds No., vehicle)
	".00436" (friction factor)
	"23.61" PSI/mi. (pressure drop, vehicle)
53.5 "D"	53.5 (drag coefficient, d_2)
.27 "E"	.27 (bed fraction, d_2)
Press "B"	1.76124 PSI/mi.
1187.4 "D"	1187.4 (drag coefficient, d_3)
.31 "E"	.31 (bed fraction, d_3)
Press "B"	.19776 PSI/mi.
"fc"	36.95 PSI/mi. (total slurry pressure drop per mile)
"fd"	1588.24 gpm (approximate)
Press "A"	1588.24
1108.5 "A"	1108.50 (30 mi. x 36.95 PSI/mi.)
.85 "A"	.85 (efficiency)
"fe"	1207.02 BHP

To clear and reset program

"fa"	resets for new case

STEP	KEY ENTRY	KEY CODE
001	f LBL A	31 25 11
	STO (i)	33 24
	f GSB 1	31 22 01
	h RTN	35 22
	f LBL 1	31 25 01
	f ISZ	31 34
	R/S	84
	h ST I	35 33
	GTO 1	22 01
010	h RTN	35 22
	f LBL B	31 25 12
	RCL 6	34 06
	1	01
	CHS	42
	+	61
	RCL 0	34 00
	1	01
	2	02
	÷	81
020	x	71
	RCL 1	34 01
	x	71
	RCL 2	34 02
	g x^2	32 54
	÷	81
	RCL 3	34 03
	f $\sqrt{x}$	31 54
	÷	81
	1	01
030	.	83
	5	05
	h y^x	35 63
	RCL 4	34 04
	1	01
	0	00
	0	00
	÷	81
	x	71
	RCL 5	34 05
040	x	71
	8	08
	2	02
	x	71
	STO A	33 11
	g GSBf b	32 22 12
	h RTN	35 22
	f LBL C	31 25 13
	RCL 2	34 02
	RCL 0	34 00
050	1	01
	2	02
	÷	81
	x	71
	RCL 6	34 06
	1	01
	CHS	42
	+	61
	RCL 7	34 07
	1	01
060	0	00
	0	00
	÷	81
	x	71
	1	01
	+	61
	6	06
	2	02
	.	83
	3	03
070	x	71
	STO E	33 15
	x	71
	RCL 8	34 08
	÷	81
	6	06
	.	83
	7	07
	2	02
	EEX	43
080	CHS	42
	4	04
	ENTER↑	41
	h R↓	35 53
	÷	81
	h PAUSE	35 72
	7	07
	7	07
	4	04
	2	02
090	÷	81
	h 1/x	35 63
	.	83
	3	03
	5	05
	5	05
	h y^x	35 63
	.	83
	0	00
	0	00
100	6	06
	6	06
	2	02
	x	71
	.	83
	0	00
	0	00
	1	01
	8	08
	+	61
110	h PAUSE	35 72
	RCL E	34 15
	6	06

STEP	KEY ENTRY	KEY CODE
	2	02
	.	83
	3	03
	÷	81
	x	71
	.	83
	3	03
120	2	02
	3	03
	x	71
	RCL 2	34 02
	g x^2	32 54
	x	71
	RCL 0	34 00
	÷	81
	5	05
	2	02
130	8	08
	0	00
	x	71
	STO B	33 12
	h RTN	35 22
	f LBL D	31 25 14
	STO 3	33 03
	h RTN	35 22
	f LBL E	31 25 15
	STO 4	33 04
140	h RTN	35 22
	g LBLf e	32 25 15
	f P≷S	31 42
	RCL 0	34 00
	RCL 1	34 01
	x	71
	1	01
	7	07
	1	01
	6	06
150	÷	81
	RCL 2	34 02
	÷	81
	f P≷S	31 42
	h RTN	35 22
	g LBLf a	32 25 11
	f CL REG	31 43
	f P≷S	31 42
	f CL REG	31 43
	h RTN	35 22
160	g LBLf b	32 25 12
	RCL A	34 11
	STO +9	33 61 09
	h RTN	35 22
	g LBLf c	32 25 13
	RCL B	34 12
	RCL 9	34 09
	+	61
	h RTN	35 22
	g LBLf d	32 25 14
170	RCL 0	34 00
	g x^2	32 54
	RCL 2	34 02
	x	71
	.	83
	0	00
	1	01
	1	01
	9	09
	÷	81
180	4	04
	2	02
	x	71
	2	02
	4	04
	÷	81
	6	06
	0	00
	÷	81
	h RTN	35 22

Preliminary Slurry Pipeline Design Procedure

Given:

It is proposed that 1.3 million tons of coal be transported for 50 miles from a mine to a power plant. This tonnage is to be shipped in 360 days in continuous operation and represents the average fuel requirement of a steam generating plant of 500 megawatts capacity operating at a 66 percent plant factor. The specific gravity of the coal is 1.4, a density of 87.5 lbs./ft.^{3}. The transporting medium is water of specific gravity 1.0 and density of 62.5 lbs./ft.^{3}. It is further proposed that the transport of a 20 mesh x 0 product be studied as compared with the transport of 8 mesh x 0 product. An arbitrary solids concentration of 40% by volume is projected for this problem.

1. Determine Slurry Characteristics

Load card SP-D-001, side 1

Store parameters in this order:

KEYSTROKES	OUTPUT
1.4 "A"	1.400 (solid density)
1.0 "A"	1.000 (liquid density)
1,300,000 "A"	1,300,000.000 (Tons/yr.)
40 "A"	40.000 (arbitrarily selected C_v)
1.014 "A"	1.014 (efficiency: 365/360)

To Compute:

Press "B"	"58.12" ft.^{3}/min. (volume flowrate of solids)
	"145.32" ft.^{3}/min. (volume flowrate of slurry)
	"87.19" ft.^{3}/min. (volume flowrate of water)
	"5431.92" lbs./min. (flowrate of water)
	"72.27" lbs./ft.^{3} (density of slurry)
	"1.16" (ratio) (density of slurry)
	"48.3" % (concentration by weight)

2. Determine Settling Velocity and Drag Coefficients

Load card SP-D-003, side 1 & 2

KEYSTROKES	OUTPUT
"f CL REG"	
"CL x"	0.00

Compute: ω & C_D for particles in water

Store parameters in this order:

KEYSTROKES	OUTPUT
87.5 "A"	87.50 (slurry)
62.5 "A"	62.50 (water)
32.2 "A"	32.20 (acceleration - gravity)
841 "A"	841.00 (particle diameter - microns)
"fc" "A"	1.076×10^{-6} (kinematic viscosity of water - ft.^{2}/sec.)

To Compute:

Press "B"	"2.0" (intermediate law)
	".15" ft./sec. (settling velocity in water)
	"2.09" (drag coefficient in water)

Compute: ω & C_D for second particle diameter in water

2,380 "fb"	"2.0" (intermediate law)
	".46" ft./sec.
	".64" (drag coefficient)

Compute: ω & C_D for particles in slurry

"fa"	
"CL x"	0.00

Store parameters in this order:

87.5 "A"	87.50
72.3 "A"	72.30 (slurry)
32.2 "A"	32.20
841 "A"	841.00 (diameter in microns)
"fc" ENTER↑ 25 x	
1.16 ÷ "A"	2.319×10^{-4} (kinematic viscosity - c.s.)

To Compute:

Press "B"	"1.0" (Stokes law)
	".01235" ft./sec. (settling velocity in slurry)
	"163.4" (drag coefficient)
(8 mesh) 2,380 "fb"	"2.0" (intermediate law)
	".09" ft./sec.
	"8.4" (drag coefficient)

2,380 microns = 8 mesh = (.0937 inch) ω = .09 ft./sec.
841 microns = 20 mesh = (.0331 inch) ω = .01235 ft./sec.

3. Determine Viscosity

A limiting value of the viscosity may be obtained using the Einstein basic correlation for Newtonian fluids. Card SP-D-002.

However, the viscosity of a coal slurry increases rapidly beyond a solid weight concentration of 50%. The value at 60°F is approximately 25 centipoise @ 50% wt.

4. Determination of Flow Parameters

A. Preliminary Selection of Pipe Sizes - Approximation

KEYSTROKES	OUTPUT
"f CL REG"	
"CL x"____________	0.00
Load card SP-F-001	
1.3 "E"____________	1.30

To Compute:

(coal) 2 "D"____________	"9.5" inches (I.D.)-feasible I.D.
____________	"650" acre-ft./yr. (water)
____________	"3.79 x 10^6" U.S. gallons/yr. (water)

So line sizes to be investigated are: 8", 10", 12".

B. Determination of Transport Velocity, Critical Velocity, Transition Velocity

Store parameters in this order:

145.32 "A"____________	145.32000 (flowrate - ft.^{3}/min.)
8.0 "A"____________	8.00000 (I.D.)
1.0 "A"____________	1.00000 (F_L)
.0331 "A"____________	.03310 (particle diameter, inch)
.01235 "A"____________	.01235 (settling velocity in slurry)
1.4 "A"____________	1.40000 (solid density)

To Compute:

Press "B"____________	6.925 ft./sec. (transport velocity)
Press "C"____________	"4.14" ft./sec. (critical velocity - Durand)
____________	"1.66" ft./sec. (critical velocity - Wasp et.al.)
____________	"7.8" ft./sec. (transition velocity - Newitt et.al.)

Compute Velocity Parameters for 8 mesh particle (.0937 inches)

.0937 "fb"____________	.0937 (particle diameter - inch)
.09 "fc"____________	.0900 (settling velocity in slurry)
Press "B"____________	6.925 ft./sec. (transport velocity)

KEYSTROKES	OUTPUT
Press "C"	"4.144" ft./sec. (critical velocity - Durand)
	"1.97" ft./sec. (critical velocity - Wasp)
	"15.11" ft./sec. (transition velocity) (- Newitt)

Compute parameters for 10 inch pipeline and 8 mesh particle.

10 "fd"	10.00
Press "B"	4.432 ft./sec. (transport velocity)
Press "C"	"4.63" ft./sec. (critical velocity - Durand)
	"2.12" ft./sec. (critical velocity - Wasp)
	"16.27" ft./sec. (transition velocity - Newitt)

Compute parameters for 10 inch pipeline and 20 mesh particle.

.0331 "fb"	.0331 (particle diameter - inch)
.01235 "fc"	.01235 (particle settling velocity- ft./sec.)
Press "B"	4.43 ft./sec. (transport velocity)
Press "C"	"4.63" ft./sec. (critical velocity - Durand)
	"1.78" ft./sec. (critical velocity - Wasp)
	"8.4" ft./sec. (transition velocity - Newitt)

Compute parameters for 12 inch pipeline and 20 mesh particle.

12 "fd"	12.00
Press "B"	3.08 ft./sec. (transport velocity)
Press "C"	"5.08" ft./sec. (critical velocity - Durand)
	"1.9" ft./sec. (critical velocity - Wasp)
	"8.93" ft./sec. (transition velocity - Newitt)

Compute parameters for 12 inch pipeline and 8 mesh particle.

.0937 "fb"	.0937
.09 "fc"	.0900
Press "B"	3.07 ft./sec. (transport velocity)
Press "C"	"5.07" ft./sec. (critical velocity - Durand)
	"2.26" ft./sec. (critical velocity - Wasp)
	"17.29" ft./sec. (transition velocity - Newitt)

Pipe O.D.	Transport Velocity (ft./sec.)	Critical Velocity$_1$ (ft./sec.)	Critical Velocity$_2$ (ft./sec.)	Transition Velocity (ft./sec.)	Particle Diameter	(Slurry) Drag Coeff.	Settling Velocity (ft./sec.)
8"	6.9	4.1	1.7	7.8	20 mesh	163.4	.01235
10"	4.4	4.4	1.8	8.4	20 mesh	163.4	.01235
12"	3.1	5.1	1.9	8.9	20 mesh	163.4	.01235

Pipe O.D.	Transport Velocity (ft./sec.)	Critical Velocity$_1$ (ft./sec)	Critical Velocity$_2$ (ft./sec.)	Transition Velocity (ft./sec.)	Particle Diameter	Drag Coeff.	Settling Velocity (ft./sec.)
8"	6.9	4.1	2.00	15.1	8 mesh	8.4	.09
10"	4.4	4.6	2.12	16.27	8 mesh	8.4	.09
12"	3.1	5.07	2.26	17.29	8 mesh	8.4	.09

Discussion of the 8 Mesh Particle

The calculated transport velocity for an 8 inch nominal pipe is well above the critical velocity$_1$, and the pipeline system could be operated at this velocity without danger of choking. The flow would tend to be heterogeneous. (maximum slope: 10 to 16%)

The 10 inch nominal pipe represents a borderline case. Since there is not much difference between the transport velocity and the critical velocity$_1$, the coarser particles in the slurry may settle out in selected locations such as inclined pipeline stretches or low spots.

Discussion of the 20 Mesh Particle

Since the calculated transport velocity for the 8" line is near the transition velocity, homogeneous flow can be assumed.

The slurry flow in both the 10 inch and 12 inch line can be assumed to be heterogeneous. The 10 inch line, again is a border line case. The 12 inch line transport velocity is below the critical velocity, and hence is unacceptable.

The following pipe diameters are selected for further investigation:

20 mesh x 0 product: 8 inch - homogeneous flow
8 mesh x 0 product: 8 inch - heterogeneous flow

5. Determination of Friction Head Loss & BHP, 20 mesh x 0 product 8" line

Load Card SP-F-002, side 1 & 2
"CLEAR ALL REGISTERS"
Store parameters in this order:

KEYSTROKES	OUTPUT
6.9 "A"	6.90
8.00 "A"	8.00
1.4 "A"	1.40
.4 "A"	.40
13.7 "A"	13.70 (experimentally verified drag coeff.)
50.0 "A"	50.00
.77 "A"	.77

To Compute Water Pressure Drop:

Press "B"	"55.20" (Reynolds No. Modified)
	".00652" (PSI/ft. - water)
	".01507" (ft./ft. - water)
Press "fb"	"55.2" (Reynolds No. Modified)
	".00652" (PSI/ft. - water)
	"34.45" (PSI/mile - water)
Press "C"	39.96 PSI/mi. (homogeneous pressure drop)
Press "D"	46.45 PSI/mi. (heterogeneous pressure drop)
46.45 ENTER↑	46.45
"fe"	46.45
Press "E"	"1082.35" gpm
	"2322.50" PSI
	"1903.57" H.P. (heterogeneous)
39.96 ENTER↑	39.96
"fe"	39.96
Press "E"	"1082.35" gpm
	"1998.00" PSI
	"1637.61" BHP (homogeneous)

References

4-1 Wasp et al, "Solid-Liquid Flow, Slurry Pipeline Transportation", Trans Tech Publications, 1977.

4-2 Colorado School of Mines Research Foundation Inc. "The Transportation of Solids in Steel Pipelines", 1963.

Two-Phase Vehicle Procedure

(1) Compute pressure drop for vehicle.
(2) Compute drag coefficients in water for all particle diameters.
(3) Compute drag coefficients in slurry for all particle diameters.
(4) Compute C/C_A ratio for each particle diameter using the settling velocity in slurry.
(5) Compute vehicle and bed volume fractions using C/C_A ratios.
(6) Compute vehicle ΔPSI/mi. at new C_v.
(7) Compute bed ΔPSI/mi. for d_n particles using particle drag coefficient in water.
(8) Compute total ΔPSI/mi.
(9) Compute total ΔPSI and BHP.

Nomenclature

C/C_A - ratio of volumetric concentrations of solids at 0.08D from top to pipe center
C_D - drag coefficient
C_w - weight percent solids
C_p - specific heat
C_v - volume percent solids
d - particle diameter
D - pipe diameter
F_L - Durand pressure drop equation coefficient (approx. 1.0)
f - friction factor
f_m - friction factor for slurry
g - acceleration due to gravity
k' - coefficient for vehicle pressure drop computation
k - thermal conductivity
s - density of solid (ratio)
U* - friction velocity, $v\sqrt{f/2}$
V_D - deposition velocity
V - mean velocity of flow
β - ratio of mass transfer coefficient to momentum transfer coefficient = 1
ρ_f - density of liquid
ρ_s - density of solids
ρ_m - density of mixture (slurry)
η - coefficient of rigidity
Φ - volume fraction solids
τ_o - yield stress
μ_m - viscosity of suspension (absolute)
μ_o - viscosity of suspending medium (absolute)
ν - kinematic viscosity
ξ - Von Karman constant = .38
ω - settling velocity

5.0 Heated Oil Pipeline Parameters

5.1 Introduction

The heated oil program computes parameters that will permit the computation of a pressure drop gradient per mile.

5.2 Theoretical Formulae

$$Q_1 = M\, C_p\, \Delta T_c \qquad \text{Sensible Heat Quantity}$$

$$Q_2 = U\, A\, \Delta T_D \qquad \text{Heat Transfer Rate}$$

$$a = \frac{6.28k}{MC \ln \left(\frac{2D - R}{R}\right)}$$

$$x = \frac{1}{5{,}280\, a} \ln \frac{(t_1 - t_0)}{(t_x - t_0)}$$

$$Q_3 = \frac{2\pi KL}{\cosh^{-1} (D/R)} \times (T_p - T_a)$$

5.3 Table of k factors

Soil Conditions	k Factor
sandy soil, dry, 24 inch cover	.25 to .40
sandy soil, moist, 24 inch cover	.50 to .60
sandy soil, soaked, 24 inch cover	1.10 to 1.30
in river, 60 in. cover, 60 in. water	2.00 to 2.50
sandy soil, dry, 8 inch cover	.60 to .70
sandy soil, moist to wet, 8 in. cover	1.20 to 2.40
clay soil, dry, 24 inch cover	.20 to .30
clay soil, moist, 24 inch cover	.40 to .50
clay soil, moist to wet, 24 in. cover	.60 to .90

5.4 Friction Heat

Heat gained from interaction of fluid with the pipe flow area.

$$FH = (HHP)(2545) \quad BTU/hr/L$$

Where HHP = hydraulic horsepower required to overcome frictional resistance in line segment L.

5.5 Heat Input due to Pump Inefficiency

$$HPE = \frac{H}{778\ C_p}(1/e - 1)$$

Where H = pump differential head - ft.
C_p = heat capacity of fluid - BTU/lb/^{o}F
e = pump efficiency

5.6 Heat Input Caused by Slack Line Operation

$$HSL = \frac{E}{778\ C_p}$$

Where E = elevation difference between upstream end and the downstream end of the slack line section

HP-D-001 - Heated Oil Pipeline Parameters Computation

Input:

(1) thermal conductivity of soil - BTU/hr./ft.^{2}/°F
(2) specific gravity of oil - n (ratio)
(3) oil flowrate - bph
(4) depth of cover - inches
(5) radius of pipe (O.D.) - inches
(6) initial oil temperature - ^{o}F
(7) initial ground temperature - ^{o}F

Computes:

(1) mile post - n
(2) inlet temperature at M.P. - ^{o}F
(3) temperature difference between fluid and ground at each M.P. - ^{o}F
(4) temperature at end of M.P. (outlet) - ^{o}F
(5) temperature loss at M.P. outlet - ^{o}F
(6) average temperature at M.P. midsection - ^{o}F
(7) speicifc gravity (approximate) @ temperature per M.P. - n

STEP	INSTRUCTIONS	KEYS	OUTPUT UNITS
1.	load card HP-D-001, side 1 & 2		
2.	input overall heat transfer coeff.	A	BTU/hr./ft.^{2}/^{o}F
3.	input specific gravity of oil	A	n
4.	input oil flowrate	A	BPH
5.	input depth of cover for pipeline	A	inches
6.	input radius of pipeline based on O.D.	A	inches
7.	input initial temperature of oil	A	^{o}F
8.	input initial temperature of oil	STO A	^{o}F
9.	input soil temperature	STO C	^{o}F
10.	Compute:	B	
	(1) mile post		n
	(2) inlet temperature @ M.P.		^{o}F
	(3) $(T_1 - T_o)$ at each M.P.		^{o}F
	(4) T_2 outlet temperature @ M.P.		^{o}F
	(5) $(T_1 - T_2)$ temperature differential @ M.P.		^{o}F
	(6) average temperature at M.P. midsection		^{o}F
	(7) average specific gravity		n
11.	Discussion: Viscosity is to be obtained from ASTM Standard Viscosity - Temperature charts for liquid Petroleum Products D 341.		

STEP	INSTRUCTIONS	KEYS	OUTPUT UNITS
12.	Input viscosity and specific gravity values into pressure drop computation card for liquids and compute pressure drop per mile at each M.P.		
13.	To clear and reset program for new case	fa	

Example: HP-D-001 - Heated Oil Pipeline Parameters Computation

Given Conditions:

K = 0.55 BTU/hr./ft.^{2}/^{o}F
speicifc gravity of oil @ 200°F: .95
oil flowrate: 925 BPH
depth of cover: 24 inches
radius of pipeline O.D.: 4 inches
initial temperature: 200°F
soil temperature: 50°F

Execution: Load card HP-D-001, side 1 & 2

Store parameters in this order:

KEYSTROKES	OUTPUT
.55 "A"	.55
.95 "A"	.95
925 "A"	925.00
24 "A"	24.00
4 "A"	4.00
200 "A"	200.00
200 "STO A"	200.00
50 "STO C"	50.00

To Compute: Press "B"

M.P.	T_1	$T_1 - T_o$	T_2	ΔT	T_{avg}	s.g.
1.	200	150	192.76	7.24	196.38	.90
2.	192.76	142.76	185.87	6.89	189.32	.90
3.	185.87	135.87	179.31	6.56	182.59	.902
4.	179.31	129.31	173.07	6.24	176.19	.904
5.	173.07	123.07	167.134	5.94	170.1	.906

"CL x" stops execution at the end of M.P. #5 parameters.

"fa" clears and resets program for new case.

STEP	KEY ENTRY	KEY CODE
001	f LBL B	31 25 12
	f GSB 4	31 22 04
	f GSB 2	31 22 02
	RCL D	34 14
	f -x-	31 84
	f GSB C	31 22 13
	f GSB 3	31 22 03
	GTO B	22 12
	h RTN	35 22
010	f LBL 4	31 25 04
	h π	35 73
	2	02
	RCL 0	34 00
	x	71
	x	71
	5	05
	2	02
	8	08
	0	00
020	x	71
	2	02
	.	83
	3	03
	0	00
	3	03
	÷	81
	RCL 1	34 01
	÷	81
	RCL 2	34 02
030	÷	81
	3	03
	5	05
	0	00
	÷	81
	.	83
	5	05
	÷	81
	RCL 3	34 03
	2	02
040	x	71
	RCL 4	34 04
	-	51
	RCL 4	34 04
	÷	81
	f LOG	31 53
	÷	81
	STO 6	33 06
	2	02
	.	83
050	3	03
	0	00
	3	03
	h x≷y	35 52
	÷	81
	STO E	33 15
	h RTN	35 22
	f LBL 2	31 25 02
	2	02
	3	03
060	h ST I	35 33
	g ISZ (i)	32 34
	h RTN	35 22
	1	01
	STO D	33 14
	h RTN	35 22
	f LBL C	31 25 13
	RCL D	34 14
	RCL E	34 15
	÷	81
070	g 10^x	32 53
	RCL 5	34 05
	RCL C	34 13
	-	51
	h x≷y	35 52
	÷	81
	RCL C	34 13
	+	61
	STO B	33 12
	h RTN	35 22
080	f LBL 3	31 25 03
	RCL A	34 11
	f -x-	31 84
	RCL C	34 13
	-	51
	f -x-	31 84
	RCL B	34 12
	f -x-	31 84
	RCL A	34 11
	h x≷y	35 52
090	-	51
	f -x-	31 84
	RCL A	34 11
	RCL B	34 12
	+	61
	.	83
	5	05
	x	71
	STO 7	33 07
	f -x-	31 84
100	f GSB D	31 22 14
	RCL B	34 12
	STO A	33 11
	h RTN	35 22
	f LBL A	31 25 11
	STO (i)	33 24
	f GSB 1	31 22 01
	h RTN	35 22
	f LBL 1	31 25 01
	f ISZ	31 34
110	R/S	84
	h ST I	35 33
	GTO 1	22 01

STEP	KEY ENTRY	KEY CODE
	h RTN	35 22
	g LBLf a	32 25 11
	f CL REG	31 43
	h RTN	35 22
	f LBL D	31 25 14
	.	83
	6	06
120	7	07
	RCL 1	34 01
	-	51
	STO 9	33 09
	RCL 7	34 07
	.	83
	0	00
	0	00
	0	00
	3	03
130	3	03
	x	71
	CHS	42
	1	01
	.	83
	0	00
	0	00
	0	00
	7	07
	1	01
140	+	61
	RCL 9	34 09
	x	71
	STO 9	33 09
	RCL 1	34 01
	.	83
	9	09
	8	08
	-	51
	STO 8	33 08
150	RCL 7	34 07
	.	83
	0	00
	0	00
	0	00
	7	07
	0	00
	x	71
	CHS	42
	.	83
160	6	06
	7	07
	+	61
	RCL 8	34 08
	x	71
	STO 8	33 08
	RCL 9	34 09
	+	61
	.	83

STEP	KEY ENTRY	KEY CODE
	3	03
170	1	01
	CHS	42
	÷	81
	DSP 3	23 03
	f -x-	31 84
	h RTN	35 22
	f LBL E	31 25 15
	RCL 7	34 07
	f LN	31 52
	2	02
180	0	00
	.	83
	3	03
	5	05
	CHS	42
	x	71
	1	01
	0	00
	8	08
	.	83
190	7	07
	8	08
	+	61
	f -x-	31 84
	f P⪌S	31 42
	STO 0	33 00
	f P⪌S	31 42
	h RTN	35 22

5.7 Static Insulation Computation for Above Ground Lines

Introduction

This program computes allowable static time limits with or without insulation for above ground lines. This is of particular benefit while hydrostatic testing lines and designing for hydrate control.

Theoretical Formulae

A. Bare Pipe - Liquids

(1) $Q = WC_p \Delta t$

$$= \frac{(d^2 \pi)}{(4)} (.433)(s.g.)(C_p)(t_o - t_c)$$

$= BTU/l.f.$

(2) $q = UA \Delta T_m$

$$= (1.5)(D\pi/12) \left[\frac{(t_o - t_a) - (t_c - t_a)}{\ln(t_o - t_a/t_c - t_a)} \right]$$

$= BTU/hr/l.f.$

(3) $T = Q/q = hrs.$

B. Insulated Pipe - Liquids

$$q = \frac{2\pi k \Delta t_m}{\ln(D_2/D_1)}$$

$= BTU/hr/l.f.$

C. Bare Pipe - Gases

(1) $Q = Wc_p \Delta t$

Note: c is to be at operating temperature & pressure.

$$= (d^2\pi/4)(1/144) \left[\frac{(29)(s.g.)(P_m)}{(10.73)(460 + t_o)(Z_m)} \right] \left[C_p \ (t_o - t_c) \right]$$

$= BTU/l.f.$

(2) $q = UA \Delta T_m$ BTU/hr/l.f.

(3) $T = Q/q$ hrs.

D. Insulated Pipe - Gases

$$q = \frac{2\pi k \Delta t_m}{\ln(D_2/D_1)} \quad \text{BTU/hr/l.f.}$$

HP-D-002 - Static Insulation/Time Computation

Input:

(1) internal diameter - inches
(2) specific gravity of fluid - n
(3) specific heat of fluid - n
(4) mean operating temperature - ^{o}F
(5) critical temperature - ^{o}F
(6) ambient temperature - ^{o}F
(7) insulation thickness - inches
(8) thermal conducting of insulation - $BTU/hr/ft^2/^{o}F$
(9) mean gas line operating pressure - gas line only - psia
(10) mean compressibility factor - gas line only - n

Compute:

(1) fluid sensible heat requirement
(2) heat transfer rate - bare pipe
(3) heat transfer rate - insulated pipe
(4) time required to reach critical temperature for bare pipe
(5) time required to reach critical temperature for insulated pipe

Option: Input different U value for bare pipe computation.

STEP	INSTRUCTIONS	KEYS	OUTPUT UNITS
1.	load card HP-D-002, side 1 & 2		
2.	input I.D.	A	inches
3.	input s.g.	A	n
4.	input C_p	A	n
5.	input mean operating temperature	A	^{o}F
6.	input critical temperature	A	^{o}F
7.	input ambient temperature	A	^{o}F
8.	input insulation thickness	A	inches
9.	input thermal conductivity of insulation	A	$BTU/hr/ft^2/^{o}F$
10.	input gas line operating pressure	A	psia
11.	input mean compressibility factor (Z)	A	n
12.	compute time to reach critical temperature for a liquid line-bare	B	hrs.
	Option: to change U from 1.5 $BTU/hr/ft^2/^{o}F$ to another value for liquids bare pipe only. Input U value	fb	hrs.
13.	compute time to reach critical temp. for a liquids line-insulated	C	hrs.
14.	compute time to reach critical temp. for a gas line - bare	D	hrs.
15.	compute time to reach critical temp. for a gas line - insulated	E	hrs.

Discussion:
(1) the C_p of gases varies with pressure & temperature

(2) typical k values are as follows:

urethane foam:	.11	BTU/hr/ft^2/^{o}F/inch
polystyrene:	.27	"
concrete:	12.00	"

$C = BTU/(hr)(ft^2)^oF = K/T$ Where T = insulation thickness in inches

Example #1: HP-D-002 - Static Insulation Computation

Given Conditions:

A 48" x .562 W.T. above ground line is to undergo hydrostatic testing in the following environment:

Water Temperature in line:	52°F
Ambient Air Temperature:	15°F
Critical Temperature:	32°F (freezing)

Compute:

(1) In how many hours will the water in the bare pipe freeze?
(2) Assuming 4 inches of urethane insulation, k = .14 BTU/hr/ft^2/^{o}F/in., in how many hours will the water freeze?

Note: $C = .14/4" = .035$ BTU/hr/ft^2/^{o}F

Execution: Load card HP-D-002, side 1 & 2

Store parameters in this order:

KEYSTROKES	OUTPUT
46.876 "A"	46.876
1 "A"	1.000
1 "A"	1.000
52 "A"	52.000
32 "A"	32.000
15 "A"	15.000
4 "A"	4.000
.035 "A"	.035
0 "A"	0.000
0 "A"	0.000

To Compute:

	Press B	31.6 hrs. for water in bare pipe to freeze
Option:	3.0 "ENTER"	3.0
	"fb"	15.79 hrs.
	Press C	417 hrs. for water in pipe with 4" insulation to freeze

Example #2:

Given Conditions:

A natural gas 4" line operates at a mean pressure of 1000 psi under these additional conditions:

s.g.: .7
C_p: .65
t_o: 78°F
t_c: 32°F
t_a: -15°F
t: 4 inches of insulation
k: .14 BTU/hr/ft^2/°F/in.
P_m: 1000 psi
Z: .92

Compute:

(1) Free water freezing and hydrate formation could occur at a critical temperature of 32°F. How long can the bare line remain free of ice?

(2) Same as (1) above for an insulated line with 4" of insulation.

Execution:

Store parameters in this order:

KEYSTROKES	OUTPUT
3.688 "A"	3.688
.7 "A"	.700
.65 "A"	.650
78 "A"	78.000
32 "A"	32.000
-15 "A"	-15.000
4 "A"	4.000
.035 "A"	.035
1000 "A"	1000.000
.92 "A"	.920

To Compute:

Press D ______ .087 hrs. or about 5 minutes to reach 32°F for a bare pipe

Press E ______ .7 hrs. or about 42 minutes to reach 32°F for a pipe with 4" insulation

STEP	KEY ENTRY	KEY CODE
001	f LBL A	31 25 11
	STO (i)	33 24
	f GSB 1	31 22 01
	h RTN	35 22
	f LBL 1	31 25 01
	f ISZ	31 34
	R/S	84
	h ST I	35 33
	GTO 1	22 01
010	h RTN	35 22
	f LBL 2	31 25 02
	RCL 0	34 00
	g x^2	32 54
	h π	35 73
	x	71
	4	04
	÷	81
	.	83
	4	04
020	3	03
	3	03
	x	71
	RCL 1	34 01
	x	71
	RCL 2	34 02
	x	71
	STO A	33 11
	RCL 3	34 03
	RCL 4	34 04
030	-	51
	RCL A	34 11
	x	71
	STO A	33 11
	h RTN	35 22
	f LBL 3	31 25 03
	RCL 3	34 03
	RCL 5	34 05
	-	51
	f P≷S	3142
040	STO 9	33 09
	f P≷S	31 42
	RCL 4	34 04
	RCL 5	34 05
	-	51
	STO B	33 12
	f P≷S	31 42
	RCL 9	34 09
	f P≷S	31 42
	h x≷y	35 52
050	÷	81
	f LN	31 52
	STO C	33 13
	f P≷S	31 42
	RCL 9	34 09
	f P≷S	31 42
	RCL B	34 12
	CHS	42
	+	61
	RCL C	34 13
060	÷	81
	STO B	33 12
	h RTN	35 22
	f LBL 4	31 25 04
	RCL B	34 12
	RCL 0	34 00
	x	71
	h π	35 73
	x	71
	1	01
070	2	02
	÷	81
	1	01
	.	83
	5	05
	x	71
	STO C	33 13
	h RTN	35 22
	f LBL B	31 25 12
	f GSB 2	31 22 02
080	f GSB 3	31 22 03
	f GSB 4	31 22 04
	RCL A	34 11
	RCL C	34 13
	÷	81
	h RTN	35 22
	f LBL 5	31 25 05
	RCL B	34 12
	RCL 7	34 07
	x	71
090	h π	35 73
	x	71
	2	02
	x	71
	STO D	33 14
	RCL 0	34 00
	RCL 6	34 06
	2	02
	x	71
	+	61
100	RCL 0	34 00
	÷	81
	f LN	31 52
	RCL D	34 14
	h x≷y	35 52
	÷	81
	STO D	33 14
	h RTN	35 22
	f LBL C	31 25 13
	f GSB 2	31 22 02
110	f GSB 3	31 22 03
	f GSB 5	31 22 05
	RCL A	34 11

STEP	KEY ENTRY	KEY CODE
	RCL D	34 14
	÷	81
	h RTN	35 22
	f LBL 6	31 25 06
	RCL 0	34 00
	g x^2	32 54
	h π	35 73
120	x	71
	4	04
	÷	81
	1	01
	4	04
	4	04
	÷	81
	STO E	33 15
	RCL 1	34 01
	2	02
130	9	09
	x	71
	RCL 8	34 08
	x	71
	1	01
	0	00
	.	83
	7	07
	3	03
	÷	81
140	RCL 9	34 09
	÷	81
	f P≷S	31 42
	STO 8	33 08
	f P≷S	31 42
	RCL 3	34 03
	4	04
	6	06
	0	00
	+	61
150	f P≷S	31 42
	RCL 8	34 08
	f P≷S	31 42
	h x≷y	35 52
	÷	81
	f P≷S	31 42
	STO 8	33 08
	f P≷S	31 42
	RCL 2	34 02
	RCL E	34 15
160	x	71
	STO E	33 15
	RCL 3	34 03
	RCL 4	34 04
	-	51
	RCL E	34 15
	x	71
	f P≷S	31 42
	RCL 8	34 08

STEP	KEY ENTRY	KEY CODE
	f P≷S	31 42
170	x	71
	STO E	33 15
	h RTN	35 22
	f LBL D	31 25 14
	f GSB 3	31 22 03
	f GSB 4	31 22 04
	f GSB 6	31 22 06
	RCL E	34 15
	RCL C	34 13
	÷	81
180	f P≷S	31 42
	STO 7	33 07
	f P≷S	31 42
	h RTN	35 22
	f GSB 6	31 22 06
	h RTN	35 22
	f LBL E	31 25 15
	f GSB 3	31 22 03
	f GSB 5	31 22 05
	f GSB 6	31 22 06
190	RCL E	34 15
	RCL D	34 14
	÷	81
	f P≷S	31 42
	STO 6	33 06
	f P≷S	31 42
	h RTN	35 22
	g LBLf b	32 25 12
	f P≷S	31 42
	STO 5	33 05
200	f P≷S	31 42
	f GSB 2	31 22 02
	f GSB 3	31 22 03
	f GSB 4	31 22 04
	1	01
	.	83
	5	05
	÷	81
	f P≷S	31 42
	RCL 5	34 05
210	f P≷S	31 42
	x	71
	RCL A	34 11
	h x≷y	35 52
	÷	81
	h RTN	35 22

5.8 Dynamic Heat Transfer Computation for Above/Below Ground Systems

Introduction

This program evaluates all the parameters associated with the thermal insulation of flow lines. The system is designed specially for gas flow lines.

HP-D-003, A & B - Dynamic Insulation Computation

General Procedure - Load card HP-D-003A, side 1 & 2

STEP	INSTRUCTIONS	KEY
1.	store parameters	A
2.	store insulation thickness	D
3.	compute mass flowrate and C_p	B
4.	Options:	
	Buried	
	(A) input 1	R/S
	(B) Insulated-input 1	R/S
	Bare-input 0	R/S
	Above Ground	
	(A) input 0	R/S
	(B) Insulated-input 1	R/S
	Bare-input 0	R/S

- (input 1 "R/S", input 1 "R/S"), buried and insulated, (1,1)
- (input 1 "R/S", input 0 "R/S"), buried and bare, (1,0)
- (input 0 "R/S", input 1 "R/S"), above ground and insulated, (0,1)
- (input 0 "R/S", input 0 "R/S"), above ground and bare, (0,0)

STEP	INSTRUCTIONS	KEY
5.	load card HP-D-003B, side 1 & 2	
6.	store ground temperature	E
	store/compute T(initial)	A
	store/compute T(final)	B
	store/compute length of line in ft.	C
7.	store air temperature	STO 9 (above ground & bare only)
	store/compute T(initial)	fa (above ground & bare only)
	store/compute T(final)	fb (above ground & bare only)
	store length of line in feet	C (above ground & bare only)

STORAGE REGISTER CONTENTS

0 - MMCFD - gas

1 - s.g. gas

2 - BPD oil/condensate

3 - s.g. oil/condensate

4. - BPD water

5 - O.D. - inches

6 - K_{ins} - BTU/(hr.)(ft.^{2})(^{o}F/ft.)

7 - K_{soil} - BTU/(hr.)(ft.^{2})(^{o}F/ft.)

8 - H - inches (surface to pipeline center line)

9 - air temperature - ^{o}F (above ground - bare)

0 - average pressure - psia

1S - gas - lb./hr.

2S - oil - lb./hr.

3S - water - lb./hr.

4S - C_p gas

5S - C_p oil

6S - R_2

7S - R_1

8S - $\Sigma R_2 Mc_p$, ΣR_1, Mc_p, $\Sigma R_3 Mc_p$ - A.G.

9S - CLR

A - T (initial) - ^{o}F

B - T (final) - ^{o}F

C - length of line - feet

D - insulation thickness - inches

E - ground temperature - ^{o}F (air temperature for above ground and insulated only)

HP-D-003, A & B - Heat Transfer Program for Pipelines Above/Below Ground

STEP	INSTRUCTIONS	KEYS	OUTPUT UNITS
1.	load card HP-D-003A, side 1 & 2		
2.	(a) input gas volume in MMCFD	A	n MMCFD
	(b) input gas specific gravity	A	n
	(c) input oil/condensate volume	A	n BPD
	(d) input oil condensate s.g.	A	n
	(e) input water volume BPD	A	n BPD
	(f) input pipe steel O.D.	A	n inches
	(g) input thermal conductivity of insulation	A	n BTU/(hr)(ft^2)(^oF/ft)
	(h) input thermal conductivity of soil	A	n "
	(i) input distance from ground surface to pipe center line	A	n inches
	(j) input air temperature	A	n psia
	(k) input average operating pressure	A	n psia
3.	check input	h RCL I	"11" proceed to step 4
4.	enter an insulation thickness for the system (ie. 0 for bare pipe)	D	n inches
5.	compute mass flowrates and specific heats	B	0 (output automatically resets to 0)
6.	Is pipeline buried?		
	(A) YES	"1" R/S	"1"
	(1) Is pipeline insulated?		
	(a) YES		
	Compute $\Sigma R_2 Mc_p$	"1" R/S	n
	(b) NO		
	Compute $\Sigma R_1 Mc_p$	"0" R/S	n
	(2) load card HP-D-003B, side 1 & 2		
	input T_g	E	^oF
	input/compute T_1	A	^oF
	input/compute T_2	B	^oF
	input/compute L	C	ft.
	(B) NO	"0" R/S	"0"
	(1) Is pipeline insulated?		
	(a) YES		
	-Compute $\Sigma R_3 Mc_p$	"1" R/S	n
	-Load card HP-D-003B, side 1 & 2		
	input T_{air}	E	^oF
	input/compute T_1	A	^oF
	input/compute T_2	B	^oF
	input/compute L	C	ft.
	(b) NO		
	-Compute X	"0" R/S	"0"
	-Load card HP-D-003B, side 1 & 2		

STEP	INSTRUCTIONS	KEYS	OUTPUT UNITS
	Option:		
	Input new T_{air}	STO 9	^oF
	input/compute T_1	fa	^oF
	input/compute T_2	fb	^oF
	input L	C	ft.

Example: HP-D-003, A & B

Given Conditions:

A 4" gas line is to be evaluated for thermal insulation requirements using the following data:

Q: 2.5 MMCFD
s.g.: .718
condensate: 200 BPD
s.g. condensate: .58
water: .34 BPD
O.D.: 4"
K insulation: .0108 BTU/(hr)(ft^2)/(^oF)(ft)
K soil: .65 BTU/(hr)(ft^2)/(^oF)(ft)
depth of burial: 36"
air temperature: 40^oF
average pressure: 1000 psi
insulation thickness: 4.0"
initial gas temperature: 82^oF
length: 10,667 feet
ground temperature: 25^oF

Compute Variables for: (1) below ground & insulated
(2) below ground & bare
(3) above ground & insulated
(4) above ground & bare

Execution: Load card HP-D-003A, side 1 & 2

(1) Below ground & insulated

Store parameters in this order:

KEYSTROKES	OUTPUT
2.5 "A"	2.5000
.718 "A"	.7180
200 "A"	200.0000
.58 "A"	.5800
.34 "A"	.3400
4 "A"	4.0000
.0108 "A"	.0108
.65 "A"	.6500
36 "A"	36.0000
40 "A"	40.0000

KEYSTROKES	OUTPUT	
1000 "A"	1000.0	
4.0 "D"	4.0	
h "RCL I"	11.0	"Check!"

Compute C_p & mass flowrate:

Press B__________0.0 (output automatically resets to 0)

Press "f w/DATA", insert blank data card and record all data on side 1 & 2

Compute variables for below ground/insulated line

1 "R/S"	1.0
1 "R/S"	79,298.04

Load card HP-D-003B, side 1 & 2____79,298.04

25 "E"	25
82 "A"	82
10,667 "C"	10,667
Press B	74.83

Note: 3 of the above variables, T_1, T_2 & L, may be varied while holding the insulation thickness constant, ie.

	25 "E"	25
	100 "A"	100
	5,280 "C"	5,280
Compute T_2	Press B	95.17
Further:	29 "E"	29
	5,280 "C"	5,280
	95.17 "B"	95.17
Compute T_1	Press A	99.73
Further:	29 "E"	29
	95.17 "B"	95.17
	99.73 "A"	99.73
Compute L	Press C	5,284.6

Execution (2): Below ground & bare

KEYSTROKES OUTPUT

Load card HP-D-003A, side 1 & 2

Load data card side 1 & 2 from execution (1)

"RCL 6"	.0108	
0 "STO 6"	0.0	
0 "D"	0.0	
h "RCL I"	11.0	Check!

Press B__________0.00

KEYSTROKES	OUTPUT
1 "R/S"	1.0
0 "R/S"	1794.89

Load card HP-D-003 B, side 1 & 2

	KEYSTROKES	OUTPUT
	25 "E"	25
	82 "A"	82
	10,667 "C"	10,667
Compute T_2	Press B	25 (approaches ground temperature)

Note: data card is for convenience only, and all "A" keystrokes may be inputed again for each case if desired. Also, T_1, T_2, L & T_g may be varied as in execution (1).

Execution (3): above ground & insulated

KEYSTROKES	OUTPUT

Load card HP-D-003 A, side 1 & 2
Load data card side 1 & 2 from execution (1)

	KEYSTROKES	OUTPUT
	.0108 "STO 6"	.0108
	4 "D"	4.0
	h "RCL I"	11.0 Check!
Computes C_p & mass flowrates & clears registers to 0	Press B	0.0
	0 "R/S"	0.0
	1 "R/S"	.65

Load card HP-D-003 B, side 1 & 2

	KEYSTROKES	OUTPUT
	25 "E"	25
	82 "A"	82
	10,667 "C"	10,667
Compute T_2	Press B	74.58

Execution (4): Above ground & bare

KEYSTROKES	OUTPUT

Load Card HP-D-003A, side 1 & 2
Load data card side 1 & 2 from execution 1

	KEYSTROKES	OUTPUT
	0 "STO 6"	0.0
	4 "D"	4.0
	"B"	0.0
	0 "R/S"	0.0
	0 "R/S"	0.0
Load card HP-D-003B, side 1 & 2		
	25 "STO 9"	25
	82 "fa"	82
	10,667 "C"	10,667
Compute T_2	Press "fb"	25°F (approaches air temperature)

STEP	KEY ENTRY	KEY CODE
001	f LBL A	31 25 11
	STO (i)	33 24
	f GSB 1	31 22 01
	h RTN	35 22
	f LBL 1	31 25 01
	f ISZ	31 34
	R/S	84
	h ST I	35 33
	GTO 1	22 01
010	h RTN	35 22
	f LBL 2	31 25 02
	RCL 1	34 01
	3	03
	1	01
	2	02
	0	00
	x	71
	RCL 0	34 00
	x	71
020	f P≷S	31 42
	STO 1	33 01
	f P≷S	31 42
	RCL 2	34 02
	RCL 3	34 03
	x	71
	1	01
	4	04
	.	83
	6	06
030	1	01
	x	71
	f P≷S	31 42
	STO 2	33 02
	f P≷S	31 42
	RCL 4	34 04
	1	01
	4	04
	.	83
	6	06
040	1	01
	x	71
	f P≷S	31 42
	STO 3	33 03
	f P≷S	31 42
	h RTN	35 22
	f LBL 3	31 25 03
	f P≷S	31 42
	RCL 0	34 00
	.	83
050	0	00
	0	00
	0	00
	1	01
	3	03
	x	71
	CHS	42
	.	83
	6	06
	5	05
060	+	61
	h ST I	35 33
	RCL 0	34 00
	.	83
	0	00
	0	00
	0	00
	4	04
	x	71
	.	83
070	2	02
	CHS	42
	+	61
	f P≷S	31 42
	RCL 1	34 01
	x	71
	h RCL I	35 34
	+	61
	f P≷S	31 42
	STO 4	33 04
080	f P≷S	31 42
	RCL 3	34 03
	.	83
	3	03
	8	08
	x	71
	CHS	42
	.	83
	8	08
	1	01
090	+	61
	f P≷S	31 42
	STO 5	33 05
	f P≷S	31 42
	CL X	44
	R/S	84
	f ≠0	31 61
	GTO 6	22 06
	GTO 7	22 07
	f LBL 6	31 25 06
100	R/S	84
	f ≠0	31 61
	GTO 4	22 04
	RCL 5	34 05
	RCL 8	34 08
	h x≷y	35 52
	÷	81
	4	04
	x	71
	f LN	31 52
110	2	02
	÷	81
	h π	35 73

STEP	KEY ENTRY	KEY CODE
	÷	81
	1	01
	.	83
	5	05
	÷	81
	f P≷S	31 42
	STO 7	33 07
120	RCL 1	34 01
	RCL 4	34 04
	x	71
	STO 8	33 08
	RCL 2	34 02
	RCL 5	34 05
	x	71
	RCL 8	34 08
	+	61
	STO 8	33 08
130	RCL 3	34 03
	RCL 8	34 08
	+	61
	STO 8	33 08
	RCL 7	34 07
	x	71
	STO 8	33 08
	f P≷S	31 42
	h RTN	35 22
	f LBL 8	31 25 08
140	RCL 7	34 07
	STO E	33 15
	3	03
	0	00
	STO 7	33 07
	f GSB 4	31 22 04
	RCL E	34 15
	STO 7	33 07
	h RTN	35 22
	f LBL 4	31 25 04
150	RCL D	34 14
	2	02
	x	71
	RCL 5	34 05
	+	61
	h ST I	35 33
	RCL 8	34 08
	h x≷y	35 52
	÷	81
	4	04
160	x	71
	f LN	31 52
	2	02
	÷	81
	h π	35 73
	÷	81
	RCL 7	34 07
	÷	81
	f P≷S	31 42

STEP	KEY ENTRY	KEY CODE
	STO 6	33 06
170	f P≷S	31 42
	h RCL I	35 34
	RCL 5	34 05
	÷	81
	f LN	31 52
	2	02
	÷	81
	h π	35 73
	÷	81
	RCL 6	34 06
180	÷	81
	f P≷S	31 42
	RCL 6	34 06
	+	61
	STO 6	33 06
	f P≷S	31 42
	GTO 5	22 05
	f LBL 5	31 25 05
	f P≷S	31 42
	RCL 1	34 01
190	RCL 4	34 04
	x	71
	STO 8	33 08
	RCL 2	34 02
	RCL 5	34 05
	x	71
	RCL 8	34 08
	+	61
	STO 8	33 08
	RCL 3	34 03
200	RCL 8	34 08
	+	61
	STO 8	33 08
	RCL 6	34 06
	x	71
	STO 8	33 08
	f P≷S	31 42
	h RTN	35 22
	f LBL 7	31 25 07
	R/S	84
210	f x≠0	31 61
	GTO 8	22 08
	h RTN	35 22
	f LBL B	31 25 12
	f GSB 2	31 22 02
	f GSB 3	31 22 03
	h RTN	35 22
	f LBL D	31 25 14
	STO D	33 14
	h RTN	35 22

STEP	KEY ENTRY	KEY CODE
001	f LBL A	31 25 11
	STO A	33 11
	h F? 3	35 71 03
	h RTN	35 22
	RCL B	34 12
	RCL E	34 15
	CHS	42
	+	61
	h ST I	35 33
010	RCL C	34 13
	f P≷S	31 42
	RCL 8	34 08
	f P≷S	31 42
	÷	81
	CHS	42
	g e^x	32 52
	h RCL I	35 34
	h x≷y	35 52
	÷	81
020	RCL E	34 15
	+	61
	STO A	33 11
	f GSB 7	31 22 07
	RCL A	34 11
	h RTN	35 22
	f LBL B	31 25 12
	STO B	33 12
	h F? 3	35 71 03
	h RTN	35 22
030	RCL A	34 11
	RCL E	34 15
	CHS	42
	+	61
	h ST I	35 33
	f P≷S	31 42
	RCL 8	34 08
	f P≷S	31 42
	RCL C	34 13
	h x≷y	35 52
040	÷	81
	CHS	42
	g e^x	32 52
	h RCL I	35 34
	x	71
	RCL E	34 15
	+	61
	STO B	33 12
	f GSB 7	31 22 07
	RCL B	34 12
050	h RTN	35 22
	f LBL C	31 25 13
	STO C	33 13
	h F? 3	35 71 03
	h RTN	35 22
	RCL B	34 12
	RCL E	34 15
	CHS	42
	+	61
	RCL A	34 11
060	RCL E	34 15
	CHS	42
	+	61
	÷	81
	f LN	31 52
	h ST I	35 33
	f P≷S	31 42
	RCL 8	34 08
	f P≷S	31 42
	x	71
070	CHS	42
	STO C	33 13
	f GSB 7	31 22 07
	RCL C	34 13
	h RTN	35 22
	f LBL E	31 25 15
	STO E	33 15
	h RTN	35 22
	g LBLf a	32 25 11
	STO A	33 11
080	h F? 3	35 71 03
	h RTN	35 22
	f GSB 9	31 22 09
	h ST I	35 33
	RCL C	34 13
	3	03
	x	71
	h π	35 73
	x	71
	RCL 5	34 05
090	x	71
	1	01
	2	02
	÷	81
	h RCL I	35 34
	÷	81
	1	01
	.	83
	0	00
	1	01
100	x	71
	CHS	42
	g e^x	32 52
	h ST I	35 33
	RCL B	34 12
	RCL 9	34 09
	CHS	42
	+	61
	h RCL I	35 34
	÷	81
110	RCL 9	34 09
	+	61
	STO A	33 11

STEP	KEY ENTRY	KEY CODE
	f GSB 7	31 22 07
	RCL A	34 11
	h RTN	35 22
	g LBLf b	32 25 12
	STO B	33 12
	h F? 3	35 71 03
	h RTN	35 22
120	f GSB 9	31 22 09
	h ST I	35 33
	h π	35 73
	3	03
	x	71
	RCL C	34 13
	x	71
	RCL 5	34 05
	x	71
	1	01
130	2	02
	÷	81
	h RCL I	35 34
	÷	81
	h ST I	35 33
	1	01
	.	83
	0	00
	1	01
	x	71
140	CHS	42
	g e^x	32 52
	h ST I	35 33
	RCL A	34 11
	RCL 9	34 09
	CHS	42
	+	61
	h RCL I	35 34
	x	71
	RCL 9	34 09
150	+	61
	STO B	33 12
	f GSB 7	31 22 07
	RCL B	34 12
	h RTN	35 22
	f LBL 9	31 25 09
	f P≷S	31 42
	RCL 1	34 01
	RCL 4	34 04
	x	71
160	STO 9	33 09
	RCL 2	34 02
	RCL 5	34 05
	x	71
	RCL 9	34 09
	+	61
	STO 9	33 09
	RCL 3	34 03
	RCL 9	34 09

STEP	KEY ENTRY	KEY CODE
	+	61
170	STO 9	33 09
	f P≷S	31 42
	h RTN	35 22
	f LBL 7	31 25 07
	0	00
	h ST I	35 33
	h RTN	35 22

References

5-1 F. Karge, "Modern Design of Oil Pipe Lines", Petroleum Engineer, May, 1945.

5-2 M. Martin, "Pipelining High Viscosity Crudes", Pipeline Industry, November, 1954.

5-3 H.W. Witter, "Hydraulic-Gradient Template is Useful Tool in Designing Hot Oil Pipelines", Oil & Gas Journal, September 17, 1956.

Nomenclature

a - coefficient
A - area - $ft.^2$
C - specific heat of product - crude oil .5 BTU/lb.°F
D - depth of cover - inches
K - heat transfer coefficient - $BTU/hr.ft.^2$-°F
M - weight - lbs.
L - length of pipeline - feet
Q_1 - heat quantity - BTU
Q_2 - heat transfer - BTU/hr.
Q_3 - heat transfer from pipe wall to soil - BTU/hr./L
R - radius of pipeline (outside diameter) - inches
t_1 - initial product temperature - °F
t_x - product temperature at x distance - °F
t_0 - ambient air temperature - °F (mean) - approximates soil temperature
t_p - pipeline wall temperature - °F
t_a - mean annual air temperature - °F
x - mile post - distance from initial station - miles

SECTION THREE

CONSTRUCTION

1.0 Pipeline Stress

1.1 Introduction

The stress program is designed to project a quick and accurate view of the stresses to be encountered in the design/construction of a pipeline system.

1.2 Theoretical Formulae

$$\text{hoop stress} = \frac{PD}{2t}$$

$$\text{longitudinal stress} = .5\ (PD/2t)$$

$$\text{long radius fitting (hoop stress)} = \frac{PD}{2t}\ \frac{(C - R/3)}{(C - R)}$$

$$\text{Poisson effect:}\quad .30\ \frac{PR}{t}$$

$$\text{longitudinal \& Poisson stress:}\quad \frac{PR}{5t}$$

$$\text{Anchor Loading:}\quad Q_t = 188.5\ A_m\ \Delta t$$

$$\text{(for bend)}:\quad Q_t = 377\ A_m\ \Delta t\ \sin 1/2\ \theta$$

$$\text{Anchor Loading:}\quad Q_t = 57.31\ (W)(\Delta T)$$

$$\text{Thermal Stress:}\quad 195\ \Delta T$$

Change in Length due to Temperature:

$$\Delta L = .000078 \times L \times \Delta T$$

WHERE:

P = operating pressure - psig
D = outside diameter - inches
t = wall thickness - inches
c = a factor, long radius fitting = 3R
R = pipe radius - inches
Δt = temperature change - °F
θ = bend angle
w = weight of pipeline - lbs./l.f.
L = length of pipeline - feet

Pipeline Beam Statics

$$L = (SZ/3W)^{.5}$$
$$d = 22.5\ WL^4/EI$$
$$f_n = 3.12/d^{.5}$$

WHERE:

S = allowable stress, PSI

Z = section modulus, $in.^3$
W = weight per unit length, lb./ft.
L = maximum span, ft.
E = Young's modulus, 29×10^6 used
I = moment of inertia, $in.^4$
d = deflection, in.
f = natural frequency

Concrete Coated Pipe Stress Parameters

$$R = EC/Sb$$

WHERE:

R = minimum radius of curvature, feet
E = modulus of elasticity of the material, PSI (use 30×10^6 PSI for steel, 3×10^6 PSI for concrete)
C = distance from the neutral axis to the extreme fiber (= 1/2 outside diameter) - feet
S_b = maximum allowable fiber stress, PSI (use 0.75 min. specified yield strength for steel, 3500 PSI for concrete of 4000 PSI compressive strength)

PL-C-001 - Pipeline Stress Computations

Input:

(1) design/operating pressure - psig
(2) outside diameter of pipeline - inches
(3) wall thickness of pipeline - inches
(4) radius of curvature for 3R fitting - inches
(5) length of exposed pipeline - feet
(6) initial temperature - ^oF
(7) final temperature - ^oF
(8) allowable bending stress - PSI
(9) concrete thickness - inches
(10) outside radius of bare pipe & concrete thickness - inches

Computes:

(1) hoop stress - PSI
(2) longitudinal stress - PSI
(3) circular bend hoop stress - PSI
(4) combined Poisson and longitudinal stress effect - PSI
(5) change in length due to temperature change - inches
(6) anchor force due to temperature change - lbs.
(7) thermal stress - PSI
(8) minimum allowable radius for catenary type bending - bare pipe @ specified bending stress - feet
(9) minimum allowable radius for machine bending per ANSI code - bare pipe - feet
(10) minimum allowable radius for bending with concrete coating @ specified concrete thickness - feet
(11) maximum span of bare pipe @ specified stress level - ft.
(12) deflection based on span length - inches
(13) natural frequency of span @ specified conditions - cycles/sec.

STEP	INSTRUCTIONS	KEYS	OUTPUT UNITS
1.	load card PL-C-001, side 1 & 2		
2.	input design/operating pressure	A	psig
3.	input outside pipe diameter	A	inches
4.	input pipe wall thickness	A	inches
5.	input radius of 3R fitting	A	inches
6.	input length of exposed unrestrained line section	A	feet
7.	input temperature at time of construction	A	^oF
8.	input maximum/minimum temperature to which pipe will be exposed	A	^oF
9.	input allowable bending stress	A	PSI
10.	input concrete wall thickness	A	inches
11.	input radius of concrete coated pipe	A	inches

STEP	INSTRUCTIONS	KEYS	OUTPUT UNITS
12.	Compute Parameters:	B	
	(1) hoop stress		PSI
	(2) longitudinal stress		PSI
	(3) circular bend stress		PSI
	(4) Poisson and longitudinal combined stress effect		PSI
	(5) change in length due to temperature change		inches
	(6) anchor force due to temperature change		lbs.
	(7) thermal stress		PSI
13.	Compute Parameters:	C	
	(1) minimum radius, bare pipe @ specified bending stress - catenary		feet
	(2) minimum radius, bare pipe - machine bending		feet
	(3) minimum radius for bending with specified concrete coating		feet
14.	Optional:		
	Input a new bending stress for bare pipe - catenary type bending	fc	PSI
	Computes:		
	(1) minimum allowable radius - catenary type bend		feet
	(2) minimum radius bending for machine application		feet
	(3) minimum radius for bending with concrete coating		feet
15.	Compute Beam Statics:	D	
	(1) hoop stress		PSI
	(2) longitudinal stress		PSI
	(3) circular bend stress		PSI
	(4) Poisson and longitudinal combined stress effect		PSI
	(5) change in length due to temperature change		inches
	(6) anchor force due to temperature change		lbs.
	(7) thermal stress		PSI
	(8) maximum span based on stress level		feet
	(9) maximum deflection based on span length		inches
	(10) natural frequency of span		cycles/sec.

Example: PL-C-001 - Pipeline Stress Computations

Given Conditions:

operating pressure: 1400 psig

O.D.: 18 inches
wall thickness: .500 inches
3R fitting: 25.5 inches
length of exposed pipeline: 250 feet
installation temperature: 75°F
maximum/minimum pipeline temperature: 25°F
allowable bending stress: 30,000 PSI
concrete coating thickness: 2 inches
radius of concrete coated pipe: 11 inches

Execution: Load card PL-C-001, side 1 & 2

Store parameters in this order:

KEYSTROKES	OUTPUT
1400 "A"	1,400.00
18 "A"	18.00
.5 "A"	.50
25.5 "A"	25.50
250 "A"	250.00
75 "A"	75.00
25 "A"	25.00
30,000 "A"	30,000.00
2 "A"	2.00
11 "A"	11.00

To Compute:

Press "B"	"23,800" PSI (hoop stress)
	"11,900" PSI (longitudinal stress)
	"31,733.33" PSI (circular bend stress)
	"4,760" PSI (Poisson and longitudinal stresses combined)
	".98" inches (change in length due to temperature change)
	"267,816.45" lbs. (anchor force due to temperature change)
	"9,750" PSI (thermal stress)
Press "C"	"725" feet (min. allowable radius for catenary type bending - bare pipe at specified bending stress)
	"687.02" feet (min. allowable radius for bending-machine applied bends - bare pipe)
	"785.71" feet (min. allowable radius for bending with concrete coating at specified coating thickness)

Optional: Input a new bending stress and recompute a minimum allowable radius for a catenary type bend for bare pipe - S_b = 35,000 PSI

To Compute:

	KEYSTROKES	OUTPUT
	35,000 "fc"	"621.43" feet (min. allowable radius)
		"687.02" feet (bare pipe - machine bend)
		"785.71" feet (concrete coated pipe)
(beam statics at bending stress of 35,000 PSI)	Press "D"	"23,800" PSI (hoop stress)
		"11,900" PSI (longitudinal stress)
		"31,733.33" PSI (circular bend stress)
		"4,760" PSI (Poisson and longitudinal stresses combined)
		".98" inches (change in length due to temperature change)
		"267,816.45" lbs. (anchor force due to temperature change)
		"120.88" feet (maximum span based on specified bending stress level)
		"14.69"inches (max. deflection based on span length)
		".81" cycles/sec. (natural frequency of span)
	"fa"	reset for new case

STEP	KEY ENTRY	KEY CODE
001	f LBL A	31 25 11
	STO (i)	33 24
	f GSB 1	31 22 01
	h RTN	35 22
	f LBL 1	31 25 01
	f ISZ	31 34
	R/S	84
	h ST I	35 33
	GTO 1	22 01
010	h RTN	35 22
	f LBL B	31 25 12
	RCL 1	34 01
	RCL 2	34 02
	2	02
	x	71
	-	51
	STO A	33 11
	RCL 0	34 00
	x	71
020	RCL 2	34 02
	2	02
	x	71
	÷	81
	h PAUSE	35 72
	.	83
	5	05
	x	71
	h PAUSE	35 72
	RCL A	34 11
030	.	83
	5	05
	x	71
	3	03
	÷	81
	CHS	42
	RCL 3	34 03
	+	61
	RCL A	34 11
	.	83
040	5	05
	x	71
	CHS	42
	RCL 3	34 03
	+	61
	÷	81
	RCL A	34 11
	.	83
	5	05
	x	71
050	x	71
	RCL 0	34 00
	x	71
	RCL 2	34 02
	÷	81
	h PAUSE	35 72
	RCL A	34 11
	.	83
	5	05
	x	71
060	RCL 0	34 00
	x	71
	5	05
	÷	81
	RCL 2	34 02
	÷	81
	h PAUSE	35 72
	RCL 4	34 04
	.	83
	0	00
070	0	00
	0	00
	0	00
	7	07
	8	08
	x	71
	RCL 5	34 05
	RCL 6	34 06
	-	51
	x	71
080	h PAUSE	35 72
	RCL 1	34 01
	g x^2	32 54
	RCL A	34 11
	g x^2	32 54
	-	51
	h π	35 73
	x	71
	4	04
	÷	81
090	3	03
	.	83
	4	04
	x	71
	STO B	33 12
	5	05
	7	07
	.	83
	3	03
	1	01
100	x	71
	RCL 5	34 05
	RCL 6	34 06
	-	51
	x	71
	h PAUSE	35 72
	1	01
	9	09
	5	05
	RCL 5	34 05
110	RCL 6	34 06
	-	51
	x	71

STEP	KEY ENTRY	KEY CODE
	h RTN	35 22
	f LBL C	31 25 13
	2	02
	9	09
	EEX	43
	6	06
	ENTER ↑	41
120	RCL 1	34 01
	.	83
	5	05
	x	71
	1	01
	2	02
	÷	81
	x	71
	RCL 7	34 07
	÷	81
130	h PAUSE	35 72
	RCL 1	34 01
	.	83
	0	00
	2	02
	6	06
	2	02
	÷	81
	h PAUSE	35 72
	3	03
140	EEX	43
	6	06
	ENTER ↑	41
	RCL 9	34 09
	1	01
	2	02
	÷	81
	x	71
	3	03
	5	05
150	0	00
	0	00
	÷	81
	h RTN	35 22
	g LBLf c	32 25 13
	STO 7	33 07
	f GSB C	31 22 13
	h RTN	35 22
	g LBLf a	32 25 11
	f CL REG	31 43
160	f P≷S	31 42
	f CL REG	31 43
	h RTN	35 22
	f LBL D	31 25 14
	f GSB B	31 22 12
	RCL 1	34 01
	4	04
	h y^x	35 63
	RCL A	34 11
	4	04
170	h y^x	35 63
	−	51
	.	83
	0	00
	4	04
	9	09
	1	01
	x	71
	STO C	33 13
	2	02
180	x	71
	RCL 1	34 01
	÷	81
	STO D	33 14
	RCL 7	34 07
	x	71
	3	03
	÷	81
	RCL B	34 12
	÷	81
190	f $\sqrt{x}$	31 54
	f -x-	31 84
	4	04
	h y^x	35 63
	RCL B	34 12
	x	71
	2	02
	2	02
	.	83
	5	05
200	x	71
	RCL C	34 13
	÷	81
	2	02
	9	09
	EEX	43
	6	06
	ENTER ↑	41
	h R↓	35 53
	÷	81
210	f -x-	31 84
	f $\sqrt{x}$	31 54
	3	03
	.	83
	1	01
	2	02
	÷	81
	h 1/x	35 62
	h RTN	35 22

References

1-1 Murphy, C.E., "Compute Temperature's Effect on Steel Pipe", Engineering Design Handbook, Pipeline Industry.

1-2 ter Borg, F.M., "Some Aspects on the Design of Submarine Pipelines", Calgary, Alberta.

1-3 Wilbur, W.F., "Analyzing Pipe Line Stresses", Engineering Design Handbook, Pipeline Industry.

1-4 Barc, W. et al, "Pipe Supports for Industrial Piping Systems", Proceedings of the American Power Conference, 1963, page 560.

2.0 Concrete Coating Parameters

2.1 Introduction

The Concrete Coating parameters program is designed to facilitate both the design and construction of a typical river/swamp crossing.

2.2 Theoretical Formulae

$$t = \left[\frac{(r_p^2 D_c - K/\pi)^{.5}}{(D_c - D_s)^{.5}} - r_p \right] \times 12$$

$$B_t = \pi (t/12 + r_p)^2 D_s$$

$$W_t = K + ((t/12 + r_p)^2 - r_p^2) \pi D_c$$

$W_{net} = W_t - B_t$ = lbs./l.f. (negative buoyancy)

Collapse Pressure = $.55E\ (t/D)^3$

where: E = modulus of elasticity $30\text{x}10^6$ psia (steel)
t = wall thickness - inches
D = outside pipe diameter

safety factor of 4.0

PL-C-002 - Pipeline Concrete Coating Parameters Computation

Input:

(1) outside pipe radius - feet
(2) density of concrete - $lbs./ft.^3$
(3) density of fluid - $lbs./ft.^3$
(4) wall thickness of bare pipe - inches

Computes:

(1) negative buoyancy - lbs.
(2) concrete thickness - inches
(3) weight of bare pipe - lbs./L.F.
(4) weight of concrete coated pipe in air - lbs./L.F.
(5) weight of concrete coating - lbs./L.F.
(6) volume of concrete - $ft.^3$/L.F.
(7) volume of concrete - $yd.^3$/L.F.
(8) volume of concrete - bags/L.F.
(9) volume of sand - $yd.^3$/L.F.
(10) volume of gravel - $yd.^3$/L.F.
(11) buoyancy of bare pipe - lbs./L.F.
(12) net buoyancy of bare pipe - lbs.
+ = floats
- = sinks
(13) collapse pressure of bare pipe - PSI

STEP	INSTRUCTIONS	KEYS	OUTPUT UNITS
1.	load card PL-C-002, side 1 & 2		
2.	input 0.00	D	0.00
3.	input 0.00	D	0.00
4.	input pipe radius	D	feet
5.	input density of concrete	D	$lb./ft.^3$
6.	input dinsity of fluid	D	$lb./ft.^3$
7.	input wall thickness of bare pipe	D	inches
8.	input Δx, set at .01	fa	.01
9.	input 1.0	A	1.0
10.	Guess a "t" (concrete thickness) & use Newton approximation to find actual value.	E	inches (t for zero negative buoyancy)
	Determination of t is followed by:		
11.	Compute:		
	(1) negative buoyancy		lbs.
	(2) concrete thickness		inches
	(3) wt. of bare pipe		lbs./L.F.
	(4) wt. of concrete coated pipe in air		lbs./L.F.
	(5) wt. of concrete per lineal foot		lbs.
	(6) volume of concrete		$ft.^3$/L.F.
	(7) volume of concrete		$yd.^3$/L.F.

STEP	INSTRUCTIONS			KEYS	OUTPUT UNITS
	(8)	bags of cement			per L.F.
	(9)	volume of sand			yd.^{3}/L.F.
	(10)	volume of gravel			yd.^{3}/L.F.
	(11)	buoyancy of bare pipe			lbs./L.F.
	(12)	net buoyancy for bare pipe			lbs.
	(13)	collapse pressure (safety factor = 4 for bare pipe)			PSI
12.	Optional:				
	A.	(1)	input a neg. buoyancy	fc	lbs.
		(2)	guess a "t"	E	inches
		(3)	computes step 11 (1-13) above		
	B.	(1)	input a new wall thickness	fd	inches
		(2)	guess a "t"	E	inches
		(3)	computes step 11 (1-13) above		
13.	To clear and reset for new case			fa	

Example: PL-C-002 - Concrete Coating Thickness and Parameters Computation

Given Conditions:

pipe radius: 1.5 feet
density of concrete: 140 lb./ft.^{3}
density of fluid: 75 lb./ft.^{3}
wall thickness of bare pipe: .406 inches

Execution: Load card PL-C-002, side 1 & 2

Store parameters in this order:

KEYSTROKES	OUTPUT
0.00 "D"	0.000 (constant - necessary input)
0.00 "D"	0.000 (constant - necessary input)
1.5 "D"	1.500
140 "D"	140.000
75 "D"	75.000
.406 "D"	.406
.01 "fa"	.010 (constant)
1.0 "A"	1.000 (constant)

To Compute:

Guess a "t": 3
(Newton approximation for concrete thickness)

KEYSTROKES	OUTPUT
3 "E"	"0.00" lbs. (negative buoyancy)
	"6.27" in. (concrete thickness)
	"154.36" lb./L.F. (wt. of bare pipe)
	"963.55" lb./L.F. (wt. of concrete coated pipe in air)
	"809.19" lbs. (wt. of concrete per L.F.)
	"5.78" $ft.^3$/L.F. (volume of concrete) 1:2:4
	"0.21" $yd.^3$/L.F. (volume of concrete)
	"1.34" (bags of cement per L.F.)
	"0.09" $yd.^3$/L.F. (volume of sand)
	"0.19" $yd.^3$/L.F. (volume of gravel)
	"530.14" lb./L.F. (buoyancy of bare pipe)
	"375.78" lbs. (net buoyancy of bare pipe)
	"23.67" PSI (collapse pressure)
"fa"	reset for new case

KEY ENTRY	KEY CODE
f LBL A	31 25 11
h ST I	35 33
h RTN	35 22
g LBLf e	32 25 15
h F? 0	35 71 00
GTO 0	22 00
h SF 0	35 51 00
1	01
h RTN	35 22
f LBL 0	31 25 00
0	00
h CF 0	35 61 00
h RTN	35 22
g LBLf a	32 25 11
h SF 1	35 51 01
STO E	33 15
h RTN	35 22
f LBL B	31 25 12
EEX	43
CHS	42
2	02
RCL E	34 15
h F? 1	35 71 01
h x≷y	35 52
h R↓	35 53
f %	31 82
f x=0	31 51
h LST x	35 82
STO C	33 13
2	02
÷	81
-	51
STO A	33 11
STO 0	33 00
f GSB (i)	31 22 24
STO D	33 14
RCL A	34 11
RCL C	34 13
+	61
STO 0	33 00
f GSB (i)	31 22 24
STO B	33 12
RCL D	34 14
-	51
RCL C	34 13
÷	81
h RTN	35 22
f LBL C	31 25 13
STO 0	33 00
f GSB (i)	31 22 24
h RTN	35 22
f LBL E	31 25 15
f FIX	31 23
f GSB B	31 22 12
RCL B	34 12
GTO 0	22 00
f LBL 6	31 25 06
RCL 0	34 00
f GSB (i)	31 22 24
STO B	33 12
f LBL 0	31 25 00
RCL A	34 11
RCL 0	34 00
STO A	33 11
-	51
RCL D	34 14
RCL B	34 12
STO D	33 14
-	51
÷	81
x	71
STO -0	33 51 00
RCL 0	34 00
h F? 0	35 71 00
h PAUSE	35 72
÷	81
f RND	31 24
f x≠0	31 61
GTO 6	22 06
RCL 0	34 00
f GSB 5	31 22 05
h RTN	35 22
f LBL D	31 25 14
STO (i)	33 24
f GSB 9	31 22 09
h RTN	35 22
f LBL 9	31 25 09
f ISZ	31 34
R/S	84
h ST I	35 33
GTO 9	22 09
h RTN	35 22
f LBL 7	31 25 07
RCL 2	34 02
2	02
4	04
x	71
STO 1	33 01
RCL 5	34 05
2	02
x	71
RCL 1	34 01
h x≷y	35 52
-	51
STO 1	33 01
h x≷y	35 52
g x^2	32 54
RCL 1	34 01
g x^2	32 54
-	51
4	04
÷	81

STEP	KEY ENTRY	KEY CODE
	h π	35 73
	x	71
	3	03
	.	83
	4	04
	x	71
	STO 1	33 01
120	h RTN	35 22
	f LBL 1	31 25 01
	f GSB 7	31 22 07
	RCL 0	34 00
	1	01
	2	02
	÷	81
	RCL 2	34 02
	+	61
	g x^2	32 54
130	STO 7	33 07
	h π	35 73
	x	71
	RCL 4	34 04
	x	71
	STO 8	33 08
	RCL 7	34 07
	RCL 2	34 02
	g x^2	32 54
	-	51
140	h π	35 73
	x	71
	RCL 3	34 03
	x	71
	RCL 1	34 01
	+	61
	STO 6	33 06
	RCL 8	34 08
	-	51
	RCL 9	34 09
150	-	51
	h RTN	35 22
	f LBL 5	31 25 05
	RCL 9	34 09
	f -x-	31 84
	RCL 0	34 00
	f -x-	31 84
	RCL 1	34 01
	f -x-	31 84
	RCL 6	34 06
160	f -x-	31 84
	h x≷y	35 52
	-	51
	f -x-	31 84
	RCL 3	34 03
	÷	81
	f -x-	31 84
	.	83
	0	00
	3	03
170	7	07
	x	71
	f -x-	31 84
	6	06
	.	83
	2	02
	8	08
	x	71
	f -x-	31 84
	.	83
180	0	00
	7	07
	x	71
	f -x-	31 84
	2	02
	x	71
	f -x-	31 84
	h π	35 73
	RCL 2	34 02
	g x^2	32 54
190	x	71
	RCL 4	34 04
	x	71
	f -x-	31 84
	RCL 1	34 01
	-	51
	f -x-	31 84
	RCL 5	34 05
	RCL 2	34 02
	2	02
200	4	04
	x	71
	÷	81
	3	03
	h y^x	35 63
	1	01
	6	06
	.	83
	5	05
	x	71
210	6	06
	g 10^x	32 53
	x	71
	h RTN	35 22
	g LBLf c	32 25 13
	STO 9	33 09
	h RTN	35 22
	g LBLf b	32 25 12
	f CL REG	31 43
	h RTN	35 22
220	g LBLf d	32 25 14
	STO 5	33 05
	h RTN	35 22

References

2-1 C.L. Almos et al, "Control of Concrete Jacket Weight is Critical", Pipeline Engineer, April 1970.

2-2 Brown, R.J., "Well Designed Concrete Coating Protects Offshore Pipelines". Oil & Gas Journal, February 26, 1973.

2-3 F.E. Hangs, "Concrete Coatings - How Thick?". Oil & Gas Journal, September 24, 1962.

2-4 "Determination of Net Buoyancy of Submerged Pipelines". Pipeline Industry, November, 1963.

Nomenclature

B_t - total buoyant force per lineal foot on concrete coated pipe - lbs./L.F.

D_c - density of concrete - lbs./ft.^{3}

D_s - density of fluid - lbs./ft.^{3}

K - weight of pipe and contents - lbs.

r - radius of bare pipe - feet

t - concrete wall thickness for a zero negative buoyancy - inches

W_{net} - net weight of concrete coated pipe in fluid - lbs. (negative buoyancy) : $W_t - B_t$

W_t - total weight of concrete coated pipe per lineal foot in air - lbs./L.F.

3.0 Hydrostatic Testing

3.1 Introduction

The objective of this program is two fold. One, estimate water filling requirements for testing a pipeline. Two, determine how temperature will affect the test pressure.

3.2 Theoretical Formulae

1. $$V = \frac{\pi d^2}{4} \times \frac{1}{144} \times 5280$$

where:

V = fill volume in cu.ft./mile
d = inside diameter in inches

2. $$\frac{\Delta V}{\Delta P} = V\left[\frac{(D)}{(Et)}\ (1 - v^2) + C\right]$$

where:

V = fill volume in cu.ft. mile
ΔV = incremental volume added per mile in $ft.^3$
ΔP = incremental pressure in PSI
D = outside diameter of pipe in inches
E = elastic modulus in PSI - 30×10^6 PSI
t = pipe wall thickness in inches
v = Poisson's ratio - .3
C = compressibility factor for water in $in.^3/in.^3/PSI$

3. $$\beta \times 10^6 = -64.268 + 17.0105t - 0.20369t^2 + 0.0016048t^3$$

where:

β = coefficient of expansion of water
t = temperature of water in °C

4. $$\Delta V = V(\beta - 2\alpha)$$

5. $$\Delta P = \frac{\beta - 2\alpha}{D/Et\ (1 - v^2) + C}$$

where:

ΔP = pressure rise in PSI for 1°C rise in temperature
α = (1.116×10^{-5})/°C (coeff. of linear expansion for steel at 15°C)
β = (16×10^{-5})/°C
C = 3.12×10^{-6}

3.3 Compressibility of Water

Temp. (°C)	Pressure (psig)	Factor C ($in.^3/in.^3/PSI$)
0	0 - 350	3.57×10^{-6}
10	0 - 350	3.4
20	0 - 350	3.34
0	350 - 720	3.51
10	350 - 720	3.35
20	350 - 720	3.24
0	1450 - 2925	3.35
10	1450 - 2925	3.14
20	1450 - 2925	3.01
50	1450 - 2925	2.89

PL-C-003 - Hydrostatic Test Computations

Input:

(1) outside pipeline diameter - inches
(2) wall thickness of pipeline - inches
(3) Poisson's ratio - .3
(4) modulus of elasticity for steel - 30 x 10^6 PSI
(5) initial test temperature - ^oC
(6) final test temperature - ^oC
(7) test pressure - psig
(8) length of test section - feet

Computes:

(1) compressibility of water @ temperature and pressure specified - x 10^{-6} in.^{3}/in.^{3}/PSI
(2) internal diameter of pipeline - inches
(3) volume of test section - ft.^{3}
(4) volume of test section - gallons
(5) differential volume for a 1 PSI rise after section is full and ready for test - ft.^{3}/PSI
(6) total added volume to achieve test pressure
- gallons (U.S.)
- BBLS
(7) differential pressure due to 1^oC temperature change - PSI/^oC
(8) differential pressure due to total temperature change - PSI

STEP	INSTRUCTIONS	KEYS	OUTPUT UNITS
1.	load card PL-C-003, side 1 & 2		
2.	input pipeline O.D.	A	inches
3.	input pipeline W.T.	A	inches
4.	input Poisson's ratio (.3)	A	.3
5.	input modulus of elasticity	A	30 x 10^6
6.	input initial test temperature	A	^oC
7.	input final test temperature	A	^oC
8.	input test pressure	A	psig
9.	input length of test section	A	feet
10.	Compute:	B	
	(1) compressibility of water at specified temperature and pressure		x10^{-6} in.^{3}/in.^{3}/PSI
	(2) I.D. of pipeline		inches
	(3) volume of test section		ft.^{3}
	(4) volume of test section		gallons
	(5) differential volume for a 1 PSI rise after section is full and ready for test		PSI

STEP	INSTRUCTIONS	KEYS	OUTPUT UNITS
	(6) total added volume to achieve test pressure		-gallons -BBLS
	(7) differential pressure due to $1^{o}C$ temperature change		$PSI/^{o}C$
	(8) differential pressure due to total temperature change		PSI
11.	To reset program for new case	f CL REC	

Example: PL-C-003 - Hydrostatic Test Computations

Given Conditions:

pipeline O.D.: 22 inches
pipeline W.T.: .321 inches
Poisson's ratio: .3
modulus of elasticity: 30 x 10^6 PSI
initial test temperature: $21^{o}C$
final test temperature: $19^{o}C$
test pressure: 1500 psig
length of test section: 105,600 feet (20 mi. x 5280)

Execution: Load card PL-C-003, side 1 & 2

Store parameters in this order:

KEYSTROKES	OUTPUT
22 "A"	22.000
.321 "A"	.321
.3 "A"	.300
30 x 10^6 "A"	30 x 10^6
21 "A"	21.000
19 "A"	19.000
1500 "A"	1,500.000
105,600 "A"	105,600.000

To Compute:

Press "B"	"3.1" (multiply this value by: 10^{-6} in.^{3}/in.^{3}/PSI)
	"21.36" inches
	"262,731.69" ft.^{3}
	"1,965,233.08" gallons (volume of test section)
	"1.36" ft.^{3}/PSI
	"15,267.698" gallons
	"363.5166" BBLS
	"35.7187" $PSI/^{o}C$
	"71.437" PSI (total differential pressure due to temperature change)
f CL REC	to reset program for new case

PL-C-003

STEP	KEY ENTRY	KEY CODE
001	f LBL A	31 25 11
	STO (i)	33 24
	f GSB 1	31 22 01
	h RTN	35 22
	f LBL 1	31 25 01
	f ISZ	31 34
	R/S	84
	h ST I	35 33
	GTO 1	22 01
010	h RTN	35 22
	f LBL B	31 25 12
	RCL 6	34 06
	3	03
	5	05
	0	00
	g $x \leq y$	32 71
	GTO 2	22 02
	3	03
	.	83
020	5	05
	5	05
	RCL 4	34 04
	RCL 5	34 05
	+	61
	.	83
	5	05
	x	71
	.	83
	0	00
030	1	01
	CHS	42
	x	71
	+	61
	STO A	33 11
	GTO 4	22 04
	f LBL 2	31 25 02
	RCL 6	34 06
	7	07
	2	02
040	0	00
	g $x \leq y$	32 71
	GTO 3	22 03
	3	03
	.	83
	5	05
	RCL 4	34 04
	RCL 5	34 05
	+	61
	.	83
050	5	05
	x	71
	.	83
	0	00
	1	01
	CHS	42
	x	71

STEP	KEY ENTRY	KEY CODE
	+	61
	STO A	33 11
	GTO 4	22 04
060	f LBL 3	31 25 03
	3	03
	.	83
	3	03
	RCL 4	34 04
	RCL 5	34 05
	+	61
	.	83
	5	05
	x	71
070	.	83
	0	00
	1	01
	CHS	42
	x	71
	+	61
	STO A	33 11
	GTO 4	22 04
	f LBL 4	31 25 04
	RCL A	34 11
080	h PAUSE	35 72
	RCL 0	34 00
	RCL 1	34 01
	2	02
	x	71
	-	51
	h PAUSE	35 72
	g x^2	32 54
	h π	35 73
	x	71
090	4	04
	÷	81
	1	01
	4	04
	4	04
	÷	81
	RCL 7	34 07
	x	71
	STO B	33 12
	h PAUSE	35 72
100	7	07
	.	83
	4	04
	8	08
	x	71
	h PAUSE	35 72
	RCL 2	34 02
	g x^2	32 54
	CHS	42
	1	01
110	+	61
	RCL 0	34 00
	x	71

STEP	KEY ENTRY	KEY CODE
	RCL 3	34 03
	÷	81
	RCL 1	34 01
	÷	81
	RCL A	34 11
	EEX	43
	6	06
120	CHS	42
	x	71
	+	61
	f P≷S	31 42
	STO 0	33 00
	f P≷S	31 42
	RCL B	34 12
	x	71
	STO C	33 13
	h PAUSE	35 72
130	7	07
	.	83
	4	04
	8	08
	0	00
	5	05
	x	71
	RCL 6	34 06
	x	71
	h PAUSE	35 72
140	4	04
	2	02
	÷	81
	h PAUSE	35 72
	RCL 4	34 04
	RCL 5	34 05
	+	61
	.	83
	5	05
	x	71
150	STO 8	33 08
	3	03
	h y^x	35 63
	.	83
	0	00
	0	00
	1	01
	6	06
	0	00
	4	04
160	8	08
	x	71
	STO 9	33 09
	RCL 8	34 08
	g x^2	32 54
	.	83
	2	02
	0	00
	3	03

STEP	KEY ENTRY	KEY CODE
	6	06
170	9	09
	x	71
	CHS	42
	STO +9	33 61 09
	RCL 8	34 08
	1	01
	7	07
	.	83
	0	00
	1	01
180	0	00
	5	05
	x	71
	STO +9	33 61 09
	6	06
	4	04
	.	83
	2	02
	6	06
	8	08
190	CHS	42
	STO +9	33 61 09
	RCL 9	34 09
	EEX	43
	6	06
	CHS	42
	x	71
	STO 9	33 09
	2	02
	.	83
200	2	02
	3	03
	2	02
	ENTER ↑	41
	5	05
	CHS	42
	g 10^x	32 53
	x	71
	CHS	42
	RCL 9	34 09
210	+	61
	f P≷S	31 42
	RCL 0	34 00
	f P≷S	31 42
	÷	81
	STO D	33 14
	h PAUSE	35 72
	RCL 4	34 04
	RCL 5	34 05
	-	51
220	RCL D	34 14
	x	71
	STO E	33 15
	h RTN	35 22

References

3-1 L.E. Brooks, "Hydrostatic Testing Procedures and Results". Fundamentals of Pipeline Construction - Petroleum Engineer. Dallas, Texas.

3-2 J.C. Gray, "How Temperature Affects Pipeline Hydrostatic Testing", Pipeline and Gas Journal, December, 1976.

4.0 Blowdown Assembly

4.1 Introduction

This program is designed to compute the riser I.D. and/or the blowdown time. Additionally, it will compute the instantaneous thrust and exhaust rate.

4.2 Theoretical Formulae

Thrust: $P = (W/g \times V_a)$ where,

P = thrust reaction - lbs.
W = discharge of gas - lb./sec.
g = acceleration due to gravity - ft./sec.^{2}
V_a = sonic velocity of gas - ft./sec.

$V_a = \sqrt{kgRT}$ where,
k = ratio of specific heats (1.23)
R = gas constant - 1544/(29)(s.g. of gas)
T = °R @ flowing conditions

$Q = D^2 P_m$ where,
D = I.D. of riser
P_m = pressure near riser - psig
Q = mcf/hr.

Joule-Thompson Effect:

$$\frac{P_m}{14.73} (°1F) = °F$$

$$T_m = \frac{0.0588\ (P)^{1/3}\ (G)^{1/2}\ (D^2)\ (L)\ (F_c)}{(d^2)\ (n)}$$

PL-C-004 - Blowdown Assembly Computations

Input:

(1) inlet pressure - psig
(2) average pressure near riser - psig
(3) gas specific gravity - n
(4) I.D. of pipe - inches
(5) length of pipeline - miles
(6) choke factor - n
(7) flow temperature - ^oF
(8) no. of risers blowing simultaneously - n

Compute/Input:

(1) blowdown time - minutes
(2) riser diameter - inches

Compute:

(1) thrust - lbs.
(2) volume flowrate - mcf/hr.
(3) exit velocity - ft./sec.

STEP	INSTRUCTIONS	KEYS	OUTPUT UNITS
1.	load card PL-C-004, side 1		
2.	input inlet pressure	A	psig
3.	input average pressure a short distance from riser	A	psig
4.	input gas specific gravity	A	n
5.	input I.D. of pipe	A	inches
6.	input length of the pipeline	A	miles
7.	input choke factor	A	n
8.	input flowing temperature	A	^oF
9.	input no. of risers blowing simultaneously	A	n
10.	input/compute: blow off time	B	minutes
11.	compute/input: internal diameter of riser	C	inches
12.	Compute thrust parameters:	D	
	(1) riser I.D.		inches
	(2) volume flowrate		mcf/hr.
	(3) exit velocity		ft./sec.
	(4) thrust		lbs.
13.	Optional:		
	(1) input new average pressure	fb	psig
	(2) compute thrust parameters	D	
	(3) input new pipeline length	fc	miles
	(4) input/compute parameters	B	
	(5) input new inlet pressure	fb	psig

STEP	INSTRUCTIONS	KEYS	OUTPUT UNITS
	(6) input/compute parameters	B	
14.	To reset for new case	fa	

Example: PL-C-004 - Blowdown Time/Riser Diameter Computation

Given Conditions:

initial line pressure: 585 psig
average pressure: 350 psig
gas specific gravity: .6
I.D. of pipeline: 30.37 inches
length of pipeline: 10.45 miles
choke factor: 3.2
flowing temperature: 60°F
no. of risers blowing simultaneously: 1
riser diameter: 8 inches

Execution: Load card PL-C-004, side 1

Store parameters in this order:

KEYSTROKES	OUTPUT
585 "A"	585.00
350 "A"	350.00
.6 "A"	.60
30.37 "A"	30.37
10.45 "A"	10.45
3.2 "A"	3.20
60 "A"	60.00
1 "A"	1.00
8 "C"	8.00

To Compute:

Press "B"__________183.58 minutes (blowdown time)

Option: Input blowdown time in minutes and compute necessary I.D. for riser.

120 "B"__________120.00
Press "C"__________9.89 inches (compute riser I.D.)

Returning to Original Example:

8 "C"__________8.00

Press "B"__________183.58 minutes (blowdown time)

To Compute Thrust Parameters:

Press "D"__________"8.00" inches (riser I.D.)

KEYSTROKES	OUTPUT
____________	"22,400" mcf/hr.
____________	"1351.86" ft./sec.
____________	"13,062.45" lbs. (instantaneous value)

Options:

(1) To compute thrust parameters:

new average pressure: 300 psig (P_m)

300 "fb"____________300.00

Press "D"____________"8.00" inches (riser I.D.)
____________"19,200" mcf/hr.
____________"1351.86" ft./sec.
____________"11,196.39" lbs.

(2) To compute blowdown time:

new pipeline length: 5 miles

5 "fc"____________5.00

8 "C"____________8.00 (input riser I.D.)

Press "B"____________87.84 minutes (blowdown time)

(3) To compute blowdown time:

new inlet pressure: 500 psig (P_1)

500 "fd"____________500.00

8 "C"____________8.00 (input riser I.D.)

Press "B"____________83.36 minutes (blowdown time)

"fa"____________resets program for new case

STEP	KEY ENTRY	KEY CODE
001	f LBL A	31 25 11
	STO (i)	33 24
	f GSB 1	31 22 01
	h RTN	35 22
	f LBL 1	31 25 01
	f ISZ	31 34
	R/S	84
	h ST I	35 33
	GTO 1	22 01
010	h RTN	35 22
	f LBL B	31 25 12
	STO 8	33 08
	h F? 3	35 71 03
	h RTN	35 22
	f GSB 2	31 22 02
	RCL 9	34 09
	g x^2	32 54
	÷	81
	RCL 7	34 07
020	÷	81
	STO 8	33 08
	STO A	33 11
	h RTN	35 22
	f LBL 2	31 25 02
	RCL 0	34 00
	3	03
	h 1/x	35 62
	h y^x	35 63
	.	83
030	0	00
	5	05
	8	08
	8	08
	x	71
	RCL 2	34 02
	f $\sqrt{x}$	31 54
	x	71
	RCL 3	34 03
	g x^2	32 54
040	x	71
	RCL 4	34 04
	x	71
	RCL 5	34 05
	x	71
	h RTN	35 22
	f LBL C	31 25 13
	STO 9	33 09
	h F? 3	35 71 03
	h RTN	35 22
050	f GSB 2	31 22 02
	RCL 7	34 07
	÷	81
	RCL 8	34 08
	÷	81
	f $\sqrt{x}$	31 54
	STO 9	33 09

STEP	KEY ENTRY	KEY CODE
	STO B	33 12
	h RTN	35 22
	f LBL D	31 25 14
060	RCL 9	34 09
	h PAUSE	35 72
	g x^2	32 54
	RCL 1	34 01
	x	71
	h PAUSE	35 72
	.	83
	0	00
	1	01
	3	03
070	8	08
	9	09
	x	71
	3	03
	2	02
	.	83
	2	02
	÷	81
	STO C	33 13
	2	02
080	1	01
	0	00
	8	08
	.	83
	6	06
	7	07
	RCL 2	34 02
	÷	81
	RCL 6	34 06
	4	04
090	6	06
	0	00
	+	61
	x	71
	f $\sqrt{x}$	31 54
	h PAUSE	35 72
	RCL C	34 13
	x	71
	h RTN	35 22
	g LBLf a	32 25 11
100	f CL REG	31 43
	h RTN	35 22
	g LBLf b	32 25 12
	STO A	33 01
	h RTN	35 22
	g LBLf c	32 25 13
	STO 4	33 04
	h RTN	35 22
	g LBLf d	32 25 14
	STO 0	33 00
110	h RTN	35 22

References

4-1 J. Oldrin, "Method Speeds Pipe Venting Time Calculations", Oil & Gas Journal, February 2, 1976.

4-2 A. Wolfensperger, "Design of Gas Pipeline Blowdowns", Pipeline Engineer, December, 1964.

Nomenclature

T_m - blowdown time, minutes
P_1 - initial line pressure, psig
G - specific gravity of gas
D - I.D. of line, inches
L - length of line blown, miles
F_c - "choke" factor

Type	F_c
Ideal Nozzle	1.0
Thruport Gate	1.6
Regular Gate	1.8
"Regular" LPV	2.0
"Venturi" LPV	3.2

d - I.D. of blowdown stack, inches
n - number of stacks blowing simultaneously
P_m - average pressure - psig (near the riser)

5.0 Construction Parameters

5.1 Introduction

This program is designed to compute approximate construction material requirements for a typical pipeline operation without extreme temperatures or profile conditions.

5.2 Theoretical Formulae

(1) Fuel Consumption:

Gasoline: (BHP)(.112)(.1) = gals./hr.
Diesel: (BHP)(.107)(.1) = gals./hr.
Diesel (generator): $\frac{(BHP)(8000)}{150{,}000}$ = gals./hr.

(2) Primer Consumption:

O.D. x 3.45 = gals/mile

(3) Coal Tar Enamel:

(O.D. x .6)(1.15) = tons/mile

(4) Asphalt:

(O.D. x .6)(1.15)(.8) = tons/mile

(5) Wrapping:

(O.D. x 15) = squares/mile

(6) Excavation Requirements:

(D)(W)(1.36) = $yd.^3$/mile

(7) Cement Requirements:

(a) $\frac{(L)(W)(D)}{4.86}$ = bags of cement (94 lbs.)

(b) 1:2:4 - Cubic Yard Requirements
  (1) 6 bags of cement
  (2) 7 gals. of water
  (3) .9 $yd.^3$ of gravel
  (4) .45 $yd.^3$ of sand

(8) Pipe Weight:

$(Dt - t^2) \times 10.68$ = lbs./l.f.

(9) Carbon Equivalent:

$$C + \frac{Mn}{4}$$

(10) Welds/Welder-hr. (After Stringer Bead)

$$N = \frac{120}{(OD)(\pi)}$$

(11) Welding Rod Requirements:

(a) (O.D. x 22.5) = lbs/mile

(b) Welding Rod/weld
(O.D.)(.5)(.25) = lbs/weld

(12) Wire Rope Strength:

$y = 83015.31\ x^{1.96}$

x = wire rope diameter - inches
y = lbs. (breaking strength)

PL-C-005 - Construction Materials Requirements Computation

Input:

(1) total gasoline engine brake horsepower - BHP
(2) total diesel engine brake horsepower - BHP
(3) total diesel generator BHP - BHP
(4) gasoline engine continuous operation - hrs/day
(5) diesel engine continuous operation - hrs/day
(6) pipeline outside diameter - inches
(7) pipeline wall thickness - inches
(8) pipeline length - miles
(9) pipeline ditch (depth) - inches
(10) pipeline ditch (width) - inches
(11) concrete structures - total volume - $ft.^3$
(12) truck load capacity - lbs.
(13) barge load capacity - lbs.
(14) river crossing wire rope diameter - inches
(15) number of welders per day on the job - n
(16) number of hours continous welding/welder - hrs.
(17) pipe carbon content - n
(18) pipe manganese content - n

Computes:

(1) gallons of gasoline/day (vehicles)
(2) gallons of diesel/day (vehicles)
(3) gallons of diesel/day (generators)
(4) gallons of primer/mile
(5) gallons of primer total
(6) coal tar enamel - tons/mile
(7) coal tar enamel - tons (total)
(8) asphalt - tons/mile
(9) asphalt - tons (total)
(10) wrapping - squares/mile
(11) wrapping - squares (total)
(12) excavation - cubic yards/mile
(13) excavation - cubic yards (total)
(14) foundation - bags of cement
(15) foundation - gallons of water
(16) foundation - cubic yards of sand
(17) foundation - cubic yards of gravel
(18) pipeline weight - lbs/l.f.
(19) pipeline weight - lbs/40' joint
(20) truck capacity - joints/truck
(21) barge capacity - joints/barge
(22) welds/hr/welder - n
(23) welds/hr - n (total)
(24) welds/day - n (total)
(25) welding speed - miles/day
(26) welding rod (butt welding) - lbs./weld
(27) welding rod (butt welding) - lbs./mile
(28) welding rod (butt welding) - lbs. (total)
(29) welding rod for stringer bead - lbs./mile
(30) welding rod for hot pass - lbs./mile
(31) welding rod for filler & final beads - lbs./mile
(32) preheating requirements - n
(33) wire rope breaking strength - lbs.

STEP	INSTRUCTIONS	KEYS	OUTPUT UNITS
1.	load card PL-C-005, side 1 & 2		
2.	input total gasoline engine BHP	A	BHP
3.	input total diesel engine BHP	A	BHP
4.	input total diesel generator BHP	A	BHP
5.	input gasoline engine continuous operation	A	hrs/day
6.	input diesel engine continous operation	A	hrs/day
7.	input pipeline O.D.	A	inches
8.	input pipeline W.T.	A	inches
9.	input pipeline length	A	miles
10.	input pipeline ditch depth	A	inches
11.	input pipeline ditch width	A	inches
12.	input concrete structures - total volume	A	$ft.^3$
13.	input truck load capacity	A	lbs.
14.	input barge load capacity	A	lbs.
15.	input river crossing wire rope diameter	A	inches
16.	input no. of welders/day on job	A	n
17.	input no. of hrs. of continuous welding/welder	A	hrs.
18.	input pipe carbon content	A	n
19.	input pipe manganese content	A	n
20.	Compute: parameters #1 - 33 as on previous page.	B	

Example: PL-C-005 - Construction Parameters Computation

Given Conditions:

The following material, equipment and personnel are available for a construction operation:

Line Pipe
166,000 L.F. - 10.75 x .159 W.T.

Equipment
9 - D7 @ 185 H.P.
2 - D8 @ 224 H.P.
2 - D6 @ 150 H.P.
1 - D4 @ 100 H.P.
2 - road patrols @ 200 H.P.
3 - Kenworth @ 200 H.P.
15 - Pickups @ 200 H.P.

Total Diesel = 3,500 H.P.
Total Gasoline = 3,000 H.P.

Personnel
12 welders, 10 hour day @ 50% efficiency (continuous welding)

Execution: Load card PL-C-005, side 1 & 2

Store parameters in this order:

KEYSTROKES	OUTPUT
3,000 "A"	3,000.000
3,500 "A"	3,500.000
250 "A"	250.000
2 "A"	2.000
5 "A"	5.000
10.75 "A"	10.750
.159 "A"	.159
31.4 "A"	31.400
36 "A"	36.000
16 "A"	16.000
50 "A"	50.000
40,000 "A"	40,000.000
3,000,000 "A"	3,000,000.000
1.5 "A"	1.500
12 "A"	12.000
5 "A"	5.000 (10 @ 50% eff. = 5 hrs.)
.25 "A"	.250
.70 "A"	.700

To Compute:

Press B	"66" gallons gasoline/day
	"175" gallons diesel/day
	"320" gallons diesel/day
	"37.09" gallons primer/mile
	"1,164.55" gallons primer/total length
Coal Tar Option	"7.42" tons/mile (coal tar enamel)
	"232.91" tons/total length (coal tar enamel)
Asphalt Option	"5.91" tons/mile (asphalt)
	"185.65" tons/total length (asphalt)
Tape Option	"161.25" sq./mile
	"5,063.25" sq./total
Excavation	"783.36" $yd.^3$/mile
	"24,597.5" $yd.^3$/total length
Concrete Structures	"10.29" bags of cement (94 lbs.)
	"11.29" gallons of water
	".72" $yd.^3$ of sand
	"1.44" $yd.^3$ of gravel
Pipe Weight	"17.98" lbs./l.f.
	"719.39" lbs./joint (40')
Truck Capacity	"55" joints/truck load
Barge Capacity	"4,170.13" joints/barge load
Welding Information	"3.55" welds/hr/welder
	"42.64" welds/hr. total
	"213.21" welds/day total

Welding Speed	____________	"1.62" miles/day
Welding Rod	____________	".67" lbs/weld (butt)
	____________	"88.71" lbs/mile (butt)
	____________	"2,785.49" lbs. total
Stringer Bead	____________	"17.74" lbs/mile
Hot Pass	____________	"17.74" lbs/mile
Filler & Final Pass	____________	"53.23" lbs/mile
* Pre-Heating requirements	______	".43"
Wire rope breaking strength	___	"183,779.48" lbs.

*Should not exceed .58

STEP	KEY ENTRY	KEY CODE
001	f LBL A	31 25 11
	STO (i)	33 24
	f GSB 1	31 22 01
	h RTN	35 22
	f LBL 1	31 25 01
	f ISZ	31 34
	R/S	84
	h ST I	35 33
	GTO 1	22 01
010	h RTN	35 22
	f LBL B	31 25 12
	RCL 0	34 00
	.	83
	0	00
	1	01
	1	01
	x	71
	RCL 3	34 03
	x	71
020	h PAUSE	35 72
	RCL 1	34 01
	.	83
	0	00
	1	01
	x	71
	RCL 4	34 04
	x	71
	h PAUSE	35 72
	RCL 2	34 02
030	1	01
	.	83
	2	02
	8	08
	x	71
	h PAUSE	35 72
	RCL 5	34 05
	3	03
	.	83
	4	04
040	5	05
	x	71
	h PAUSE	35 72
	RCL 7	34 07
	x	71
	h PAUSE	35 72
	RCL 5	34 05
	.	83
	6	06
	9	09
050	x	71
	h PAUSE	35 72
	RCL 7	34 07
	x	71
	h PAUSE	35 72
	RCL 5	34 05
		83
	5	05
	5	05
	x	71
060	h PAUSE	35 72
	RCL 7	34 07
	x	71
	h PAUSE	35 72
	RCL 5	34 05
	1	01
	5	05
	x	71
	h PAUSE	35 72
	RCL 7	34 07
070	x	71
	h PAUSE	35 72
	RCL 8	34 08
	RCL 9	34 09
	1	01
	.	83
	3	03
	6	06
	x	71
	x	71
080	h PAUSE	35 72
	RCL 7	34 07
	x	71
	h PAUSE	35 72
	f P≷S	31 42
	RCL 0	34 00
	f P≷S	31 42
	4	04
	.	83
	8	08
090	6	06
	÷	81
	h PAUSE	35 72
	1	01
	+	61
	h PAUSE	35 72
	1	01
	CHS	42
	+	61
	.	83
100	0	00
	7	07
	x	71
	h PAUSE	35 72
	2	02
	x	71
	h PAUSE	35 72
	RCL 5	34 05
	RCL 6	34 06
	x	71
110	RCL 6	34 06
	g x^2	32 54
	CHS	42

STEP	KEY ENTRY	KEY CODE
	+	61
	1	01
	0	00
	.	83
	6	06
	8	08
	x	71
120	h PAUSE	35 72
	4	04
	0	00
	x	71
	ENTER↑	41
	h PAUSE	35 72
	f P≷S	31 42
	RCL 1	34 01
	h x≷y	35 52
	÷	81
130	h PAUSE	35 72
	h R↓	35 53
	RCL 2	34 02
	h x≷y	35 52
	÷	81
	h PAUSE	35 72
	3	03
	8	08
	.	83
	2	02
140	f P≷S	31 42
	RCL 5	34 05
	÷	81
	h PAUSE	35 72
	f P≷S	31 42
	RCL 4	34 04
	x	71
	h PAUSE	35 72
	RCL 5	34 05
	x	71
150	h PAUSE	35 72
	.	83
	0	00
	0	00
	7	07
	5	05
	8	08
	x	71
	h PAUSE	35 72
	f P≷S	31 42
160	RCL 6	34 06
	1	01
	.	83
	5	05
	3	03
	h y^x	35 63
	3	03
	.	83
	9	09

STEP	KEY ENTRY	KEY CODE
	8	08
170	x	71
	STO E	33 15
	RCL 5	34 05
	h π	35 73
	x	71
	1	01
	2	02
	÷	81
	RCL E	34 15
	x	71
180	h PAUSE	35 72
	1	01
	3	03
	2	02
	x	71
	STO D	33 14
	h PAUSE	35 72
	RCL 7	34 07
	x	71
	h PAUSE	35 72
190	RCL D	34 14
	.	83
	2	02
	x	71
	h PAUSE	35 72
	RCL D	34 14
	h PAUSE	35 72
	RCL D	34 14
	.	83
	6	06
200	x	71
	h PAUSE	35 72
	f P≷S	31 42
	RCL 7	34 07
	4	04
	÷	81
	RCL 6	34 06
	+	61
	f -x-	31 84
	RCL 3	34 03
210	1	01
	.	83
	9	09
	6	06
	h y^x	35 63
	8	08
	3	03
	0	00
	1	01
	5	05
220	.	83
	3	03
	x	71
	f P≷S	31 42
	h RTN	35 22

References

5-1 "Pipe Line Rules of Thumb Notebook", Pipeline Industry, 1967.

APPENDIX

Choice of Design Procedure

The following section describes a general approach to the evaluation and design of a pipeline system.

LIQUIDS (water, crude oil, refined products)

Program Execution	Description
LP-F-001	Preliminary selection of I.D.
LP-D-002	Selection of W.T.
LP-D-001 or LP-D-009	Hydraulic gradient & station location
PL-C-005	Construction Material Requirements
PL-C-003	Hydrostatic Testing
PL-C-001	Pipeline Stress Computations
PL-E-001	Pipeline Economic Computations

GASES (natural gas)

Program Execution	Description
LP-F-003	Preliminary selection of I.D.
LP-F-002	Selection of W.T.
GP-D-001 A & B	Flowrate Computation
GP-D-003 A & B	Hydraulic Gradient Computation
PL-C-005	Construction Material Requirements
PL-C-003	Hydrostatic Testing
PL-C-001	Pipeline Stress
PL-C-004	Blowdown
Economics	

SLURRY & TWO-PHASE FLOW

General Procedures are described in their respective selections.

INTERNATIONAL SYSTEM OF UNITS

Length	meter (m)	inch x 0.0254
		feet x 0.3048
Mass	Kilogram (kg)	lbm x 0.4536
Volume	meter^3 (m^3)	barrel x .015899
Volume/Time	meter^3/second (m^3/s)	gallons (U.S.)/min. x .0000631
Time	seconds (s)	
Temperature	degrees Kelvin (K)	$^oK = {^oR} \times 5/9$
		$^oK = {^oC} + 273.15$
		$^oK = ({^oF} + 459.67)(5/9)$
Force	Newton (N) ($N = m.kg.s^{-2}$)	$lb_f \times 4.448$
Pressure	Pascal (Pa)	psi x 6895
	($Pa = N.m^{-2}$)	psf x 47.88
Energy	Joule (J) (J - N.m)	Btu x 1034
Power	Watts (W)	HP x 745.7
	($W = J.s^{-1}$)	Btu/sec x 1055
		Btu/min x 17.57

Subject Index

Notes